Satellite Photoelectric Sensing Technology

Jianjun Zhang · Jing Li

Satellite Photoelectric Sensing Technology

Communication, Navigation and Reconnaissance

Springer

Jianjun Zhang
China Academy of Space Technology
Beijing, China

Jing Li
Beijing Institute of Technology
Beijing, China

ISBN 978-3-030-89842-7 ISBN 978-3-030-89843-4 (eBook)
https://doi.org/10.1007/978-3-030-89843-4

This Springer imprint is published by the registered company Springer Nature Switzerland AG
The registered company address is: Gewerbestrasse 11, 6330 Cham, Switzerland

Preface

The rapid development and popularization of radio communication technologies have made the electromagnetic environment more and more complex, the use of frequency spectrum has become more and more extensive and spectrum resources have become increasingly scarce. In order to maximize the value of limited spectrum resources, it has become an inevitable requirement to develop electromagnetic spectrum situational awareness application systems and strengthen spectrum scientific management.

With the increase of radio frequency bandwidth, traditional electronic technology has been difficult to meet the requirements of these applications, and a lot of problems have arisen. It is difficult for analog electronic devices to achieve uniform high performance within one or several octave frequencies, resulting in a low dynamic range. As a result, traditional RF frequency measurement systems based on traditional microwave devices are confronted with the conflict between frequency intercept probability and frequency resolution. In order to obtain a high enough frequency resolution, it is necessary to increase the number of sampling routes, which in turn leads to the increase of SWaP in the device. At the same time, microwave links based on technical microwave devices are facing high-frequency loss, which ultimately limits the measurement bandwidth of the frequency measuring system. Applications such as electronic countermeasure and frequency hopping communication require high fidelity digital ultra-wideband signals from which target information is extracted by using DSP. The increased bandwidth worsens the quantization accuracy of ADC. Even if the analog-to-digital converter can convert high sampling rate and high-resolution analog signals, the battlefield information of the converted massive bits (e.g., 12-bit quantization of 5 GHz analog bandwidth will generate a bit rate of 60 Gb/s) cannot be processed in subsequent digital signal processing sections.

The development of photon technology has made people realize that, depending on its advantages of high bandwidth and low loss, photon technology will become a new analog signal processing platform to separate the perception and frequency functions from DSP and avoid the huge and high energy consumption of digital processing. The system using photon technology has the uniformity of frequency for tens of GHz radio frequencies. The advantages of photon technology include parallel processing, ultra-wideband tuning, long-distance transmission and anti-electromagnetic interference. The microwave frequency measurement system based on photon technology fundamentally breaks through the limitation of the bandwidth and dynamic range of electronic devices, greatly increases the frequency measurement range and reduces the SWaP of devices. At the same time, the response time of the photonic device is much higher than that of the electronic device, which improves the real-time performance of the frequency measurement system.

The purpose of this book is to meet the needs at home and abroad, making a profound study of satellite spectrum-sensing technology based on microwave photonics. The first chapter is the introduction, which discusses the problems faced by electromagnetic spectrum situation awareness; the second chapter is microwave photonics, which discusses the development of microwave photonics; the third chapter is spacecraft system, which discusses the composition of satellites; the fourth chapter is satellite spectrum sensing, which discusses the spectrum-sensing principle of space system; the fifth chapter is radio frequency channelization technology, which discusses various calculations of channelization technology; the sixth chapter is the basis of radio frequency photon channelization, which discusses the basic theory of channelized optical link; the seventh chapter is the production and function of optical frequency comb; the eighth chapter is the channelization receiving technology based on optical frequency comb; the ninth chapter is channelized link distortion compensation based on digital signal processing.

With the completion of this book, satellite photoelectric sensing technology will be innovatively proposed to make up for the shortcomings caused by the "electronic bottleneck" in cognitive radio, greatly promote the process of spectrum sensing and optimization, and finally drive the rapid development of "broadband satellite" business.

Bejing, China

Jianjun Zhang
Jing Li

Contents

Abbreviations

ADC	Analog-to-Digital Conversion
AO	Acousto-Optical
AOM	Acousto-Optical Modulator
APD	Avalanche Photodiode
BAW	Bulk-Acoustic-Wave
BPD	Balanced Photodiode
BPSK	Binary Phase-Shift Keying
CMT	Coupled-Mode Theory
CS	Carrier Suppression
CW	Continuous Wave
CWR	Cascade Weak Reflection
DCF	Dispersion Compensation Fiber
DE	Dispersion Element
DFB	Distributed Feed Back
DFT	Dispersion Fourier Transform
DSP	Digital Signal Processing
E/O	Electro-Optical Conversion
EDFA	Erbium-Doped Fiber Amplifier
FBG	Fiber Bragg Grating
FRM	Faraday Rotator Mirror
FSF	Frequency Shift Feedback
FSR	Free Spectrum Range
FTM	Frequency to Time Mapping
FWM	Four-Wave Mixing
GPS	Global Position System
GVD	Group Velocity Dispersion
IIP3	Third-Order Input Intercept Point
IMD	Intermodulation Distortion
LO	Local Oscillation
LPFG	Long Period Fiber Grating
MLL	Mode-Locked Laser

MSW	Magneto-Static-Wave
MZM	Mach–Zehnder Modulator
NF	Noise Figure
NLSE	Nonlinear Schrodinger Equation
O/E	Optical-Electro Conversion
ODL	Optical Delay Line
OEO	Optoelectronic Oscillator
OF	Optical Filter
OIP3	Third-Order Output Intercept Point
PBS	Polarization Beam Splitter
PD	Photo-Detector
PM	Phase Modulator
PPG	Programme Pulse Generatot
PS	Phase Shifter
RF	Radio Frequency
RIN	Relative Intensity Noise
RoF	Radio-over-Fiber
RTFT	Real-Time Fourier Transform
SAW	Surface-Acoustic-Wave
SC	Satellite Communication
SDR	Signal Deterioration Ratio
SFDR	Spurious Free Dynamic Range
SH	Spectrum Holes
SPM	Self-Phase Modulation
SS	Spectrum Sensing
TTD	True Time Delay
TWSTT	Two-Way Satellite Time Transfer
WDM	Wavelength Division Multiplex
XPM	Cross Phase Modulation

Chapter 1
Introduction

1.1 Problems in Electromagnetic Spectrum Situational Awareness

With the continuous development of technology, modern warfare has been developed into a high-tech war with information warfare as its core. The electromagnetic spectrum resource is the only strategic resource that can penetrate the strategic aspects of space, land and sea, and supports all joint functions. The competition for electromagnetic spectrum resources brings traditional electronic warfare to the strategic level and is crucial to the concept of warfare for future warfare. The information confrontation technology across multiple platforms uses electromagnetic reconnaissance and anti-reconnaissance, electromagnetic suppression and interference, anti-repression and anti-interference as means, and it becomes a hot topic for countries to compete for research. The US Department of Defense, the various services and their allies have placed the electromagnetic spectrum advantage in an unquestionable position. In particular, the emphasis on electromagnetic spectrum resources has never been more important in recent years. Relying on the advantages of electromagnetic spectrum reconnaissance and control provides protection for the freedom of movement of US military forces around the world. Lieutenant General Charles Richard, deputy commander of the US Strategic Command, publicly stated that all joint operations in the United States rely on asymmetric advantages in the operational domain, provided that they are free to access the electromagnetic spectrum, maneuver and master enemy movements.[1] Firstly, gaining and maintaining the advantages of electromagnetic spectrum resources can achieve the advantages of air, ground, sea, space and cyberspace.

With the continuous increase of RF bandwidth, traditional electronic technology has been difficult to meet the needs of the above applications, and many problems

[1] The 53rd International Symposium of the Old Crow Association.

J. Zhang and J. Li, *Satellite Photoelectric Sensing Technology*,
https://doi.org/10.1007/978-3-030-89843-4_1

have arisen. It is difficult for analog electronic devices to achieve uniform high performance in one or several octaves, which leads to low dynamic range, and makes traditional RF frequency measurement systems based on traditional microwave devices facing the contradiction between frequency intercept probability and frequency resolution. In order to obtain a high enough frequency resolution, it is necessary to increase the number of sampling paths, which leads to the increase of SWaP equipment. At the same time, the microwave links based on technology microwave devices face a great high-frequency loss, which fundamentally limits the measurement bandwidth of the frequency measurement system. Electronic countermeasures, frequency hopping communications and other applications require high fidelity digital UWB signals, from which target information can be extracted by using DSP. The increase of bandwidth makes the quantization accuracy of ADC worse. Even if analog-to-digital converters can achieve a high sampling rate and high-resolution analog signal conversion, the battlefield information of the converted massive bits (e.g., 12 bits quantization of 5 GHz analog bandwidth will generate 60 Gb/s bit rate) cannot be processed in the subsequent digital signal processing part [1]. The relation between analog bandwidth and effective bits of ADC is shown in Fig. 1.1.

With the development of photon technology, people realize that photon technology will become a new analog signal processing platform depending on its advantages of high bandwidth and low loss, in order to realize the separation of sensing and frequency conversion functions from DSP and avoid huge and high energy consumption of digital processing. The system using photon technology has the uniformity of frequency for tens of GHz radio frequency. The advantages of photon technology include parallel processing, ultra-wideband tuning, long-distance transmission and anti-electromagnetic interference. The microwave frequency measurement system based on photon technology fundamentally breaks through the limitation of the bandwidth and dynamic range of electronic devices, greatly increases the frequency measurement range and reduces the SWaP of the equipment. At the same time, the

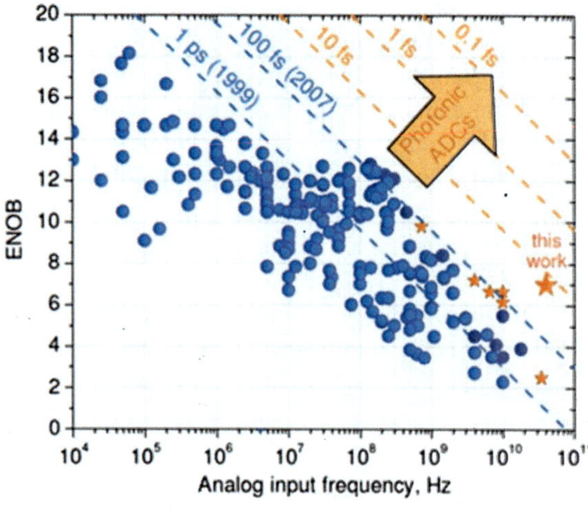

Fig. 1.1 The relation between analog bandwidth and effective bits of ADC

response time of photonic devices is much higher than that of electronic devices, which improves the real-time performance of the frequency measurement system [2].

1.2 RF Channelized Reception Based on Optoelectronic Technology

1.2.1 Research Progress of RF Channelized Receiving in Photon Technology

With the development of high-speed broadband electronic devices and analog optical link technology, broadband radio frequency channelized reception technology based on microwave photonics has been developed, which opens up a new way for broadband radio frequency reception [3–6]. The front-end module of the RF receiver transmits the radio frequency signal received by the antenna to the central station for frequency conversion demodulation and identification, realizes signal transmission and conversion between antenna and signal processor and determines the overall performance of the receiver. With the increasing complexity of the electromagnetic environment, the development of wide-band and high-frequency communication in the military and civil communications makes the processing bandwidth of the receiver and the ability of sensing and receiving strong and weak signals become the main research direction of the RF front-end. The photonic channelized reception of radio frequency is to modulate the radio frequency signal received by the antenna to the optical carrier, divide the frequency band into several narrowbands and carry out parallel channelized processing. It avoids the electronic bottleneck, thus enlarging the receiving bandwidth of the receiver and alleviating the pressure of subsequent digital signal processing. Depending on its remarkable advantages, the channelized reception technology of RF optical front-end has rapidly become a research hotspot. It has experienced the development of such technologies as surface acoustic wave (SAW) channelization technology, bulk acoustic wave (BAW) channelization technology, magnetic static wave (MSW) channelization technology and acousto-optic (AO) channelization technology.

Since 1981, R.E. Brooks, a researcher of TRW Corporation, put forward the idea of signal processing using surface acoustic wave diffraction at the seminar on microwave signal processing held by DARPA and NAVELEX, and the technology of radio frequency channelization based on SAW has been studied. In 1985, R.E. Brooks proposed a SAW channelization technique using the diffraction of an acousto-optic wave to realize Fourier transform. Based on the principle of time Fourier transform, radio frequency signals are converted into waves with different spatial angles in the process of propagation through diffraction. The principle is shown in Fig. 1.2. A phase-controlled array of SAW interdigital transducer is driven by an input radio frequency signal. The array acts as a diffraction grating to focus and distribute

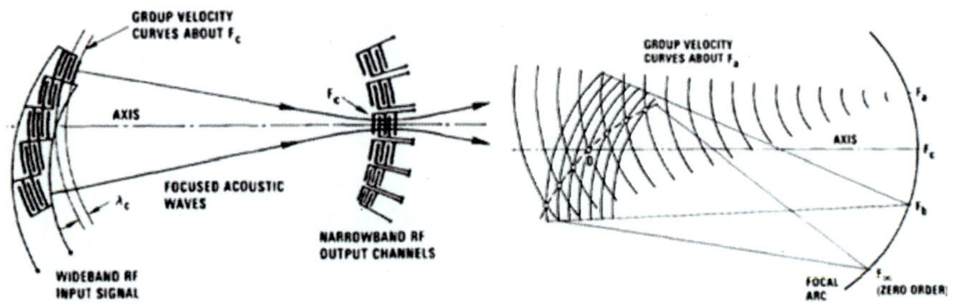

Fig. 1.2 SAW RF channelization receiving technology based on a focus phase converter

the generated SAW angle into a continuous narrowband at frequency. The scheme achieves channelized reception of 20 radio frequency signals with 5 MHz channel bandwidth in 155–255 MHz bandwidth [7].

In 1984, E.M. Alexander of the University of Maryland, funded by the Naval Aviation Command of the United States, proposed the Fabry–Perot etalon to decompose broadband radio frequency signals in multiple channels, and successfully realized spectrum detection in the range of 100–1000 MHz with a channel bandwidth of 70 MHz. However, the equalization of the channel is not satisfactory. In 1990, J.P. Powers of the Marine Graduate School in Monterrey, California, proposed a coherent acoustic-optical channelization technology for broadband radio frequency signals. The technology uses Bragg cells to physically channelize and load the spectrum of different radio frequencies. Each narrowband radio frequency signal is coupled with a local oscillator to realize Fourier transform through a dispersion device, and then coherent detection is carried out after an optical fiber array. Bragg cells use a prism to introduce dispersion to achieve the splitting function. The choice of prism focal length can set channel parameters. Each fiber represents a different channel. The scheme achieves 2.25 MHz bandwidth radio frequency reception of three channels at 75 MHz, and the coating and cladding of the optical fiber array will leak through the signal. BPSK pseudo-random sequence is used to measure the effect of channel equalization imbalance (non-ideal rectangular window shape) on data recovery. The results show that the data can be correctly restored when channel equalization is below 1 dB, and the signal quality will be seriously affected when the power jitter of frequency response decreases by more than 5 dB [8]. In 1996, Dr. Cheng Naiping, School of Electronic Information Engineering, Beijing University of Aeronautics and Astronautics, proposed an acousto-optic channelized receiver for spectrum analysis using the heterodyne reception method. The receiver can achieve a high-frequency resolution of 25 kHz, and the dynamic range of the system is greater than 50 dB [9].

In 2001, New Focus Company and Air Force Rome Laboratory of the United States, supported by DARPA, proposed a channel technology based on the free-space diffraction grating and coherent heterodyne detection. The system principle is shown in Fig. 1.3. Broadband radio frequency signals are modulated to optical carriers and input to a collimation system containing dispersive gratings and prisms. Gratings are used to separate the spectral lines of light. Prisms are used to focus

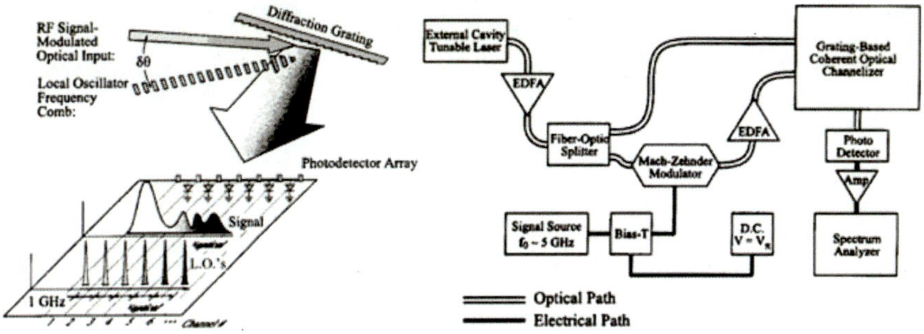

Fig. 1.3 Principle and structure of coherent heterodyne channelization receiving technology based on free-space diffraction

and collimate the input signals with gratings, gratings and photoelectric detection arrays. At the same time, multiple local oscillators are also incident on the gratings for diffraction. Each local oscillator tooth line is coupled with the corresponding narrowband signal light and input through the prism. Photoelectric conversion to their respective detectors. The instantaneous receiving bandwidth of the receiver is up to 100 GHz, the channel bandwidth is 1 GHz, all signals are down-converted to 5 GHz IF, the link gain is 90 dB, the noise index is 60 dB and the dynamic range of the channel is 101 dB Hz 2/3. The system implementation of the scheme is complicated [10]. In 2006, S.T. Winnall of Cohlear Company, Australia, and others, supported by the Ministry of Defense, studied a channelization system based on integrated Bragg grating Fabry–Perot etalon (BGFP) and integrated hybrid Fresnel lens. Broadband radio frequency is mapped to an optical carrier through electro-optic modulation and then transmitted to an integrated module through an optical fiber. Within the module, the optical phase is transversely diverged by the lens, and the divergent light waves oscillate in the standard instrument integrated by two Bragg gratings. The transmission angle and wavelength meet the corresponding resonance conditions so that they can be channeled to the subsequent photoelectric conversion module at a fixed angle to achieve channelized reception. The scheme achieves 40 channels with 1 GHz bandwidth in the receiving range of 40 GHz, and the volume of the system is reduced to 1500 mm^3, which greatly reduces the size of the system [11].

In 2001, Shane J. Strutz of the Naval Laboratory of the United States proposed the channelization technology of 8–18 GHz radio frequency reception using all-optical down-conversion. Multiple lasers and microwave local oscillators were modulated as local oscillator optical carriers. At the same time, each laser was combined into a modulated radio frequency signal through a wavelength division multiplexer, and the channel was divided by optical filtering and demultiplexing with Fabry–Perot etalon. Then coherent detection is carried out with the corresponding local oscillator light. The system experiment achieves 8–18 GHz RF bandwidth division and two channels reception, image suppression 20 dB, dynamic range 107 dB Hz 2/3, but the scheme has many devices, complex system and high cost [12]. In 2011, Camille-Sophie Brs of the University of California proposed a new channelization scheme using four-wave

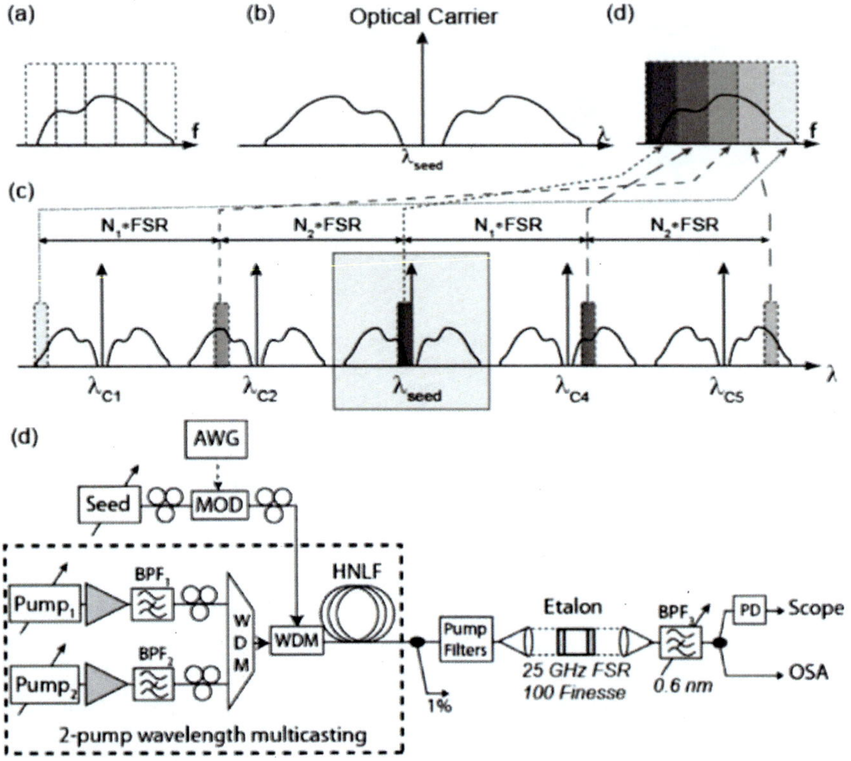

Fig. 1.4 Schematic diagram of signal-multicast microwave photonic channelization based on four-wave mixing. **a** Radio frequency signal spectrum shape. **b** The spectrum of the modulated signal under double-sided modulation. **c** Spectrum of signal multicast by four-wave mixing. **d** Channelization receives the recovered radio frequency signal. **d** A channelization experimental device based on four-wave mixing

mixing to realize signal multicast. The principle is shown in Fig. 1.4. Firstly, the seed light is used to load the RF signal, and then the wavelength of the two pumping light is set accurately to spread the seed light through the four-wave mixing effect of highly nonlinear optical fibers so as to obtain the optical carrier with multi-wavelength and realize the multicast of the modulated signal, as shown in Fig. 1.4c. Then the FP cavity is used to realize the seamless channelized cutting of broadband radio frequency by selecting appropriate FSR for periodic filtering. The scheme achieves four channels with 1 GHz bandwidth, and the extinction ratio between channels is more than 20 dB [13]. With the emergence of the optical frequency comb, channelization technology of signal multicast using optical frequency comb has been proposed. The application of optical frequency comb can not only reduce the complexity of the system but also avoid accurate calibration of wavelength or frequency between devices. In 2011, Chi Hao et al., School of Information Science and Electronic Engineering, Zhejiang University, proposed a reconfigurable channelized receiver based on optical

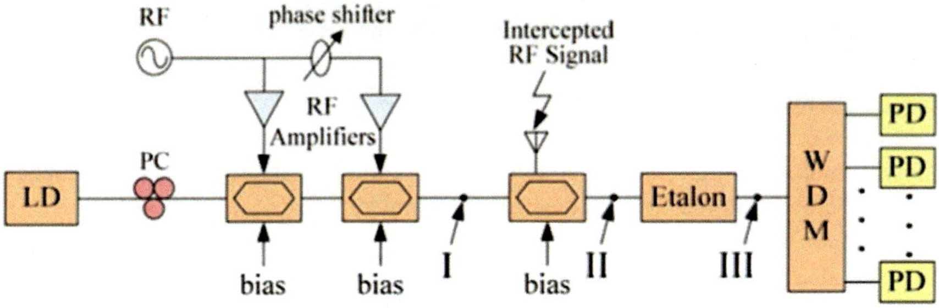

Fig. 1.5 Schematic diagram of a reconfigurable microwave photonic channelization receiver based on optical frequency comb

frequency comb. In this scheme, two cascaded Mach–Zehnder modulators (MZM) were used to generate 11 flat optical frequency combs with power jitter less than 1 dB, and then radio frequency signals were modulated to optical frequency comb by modulators to achieve signal multicast. Then, through the filter of etalon, the seamless cutting of broadband radio frequency signal can be realized by setting the frequency difference between the bandwidth of etalon and the free spectrum range of the optical frequency comb. Finally, the channel separation can be realized by WDM. The specific scheme is shown in Fig. 1.5. The scheme achieves 11 channels with a bandwidth of 1 or 1.5 GHz in the range of 0.5–11.5 or 0.75–17.25 GHz. Compared with the four-wave mixing scheme, it has prominent advantages, adjustable spectrum reconstruction, flexible and simple operation, and simplifies the structure of the system [14].

With the development of microwave photonics, the research of channelized broadband radio frequency reception using photon technology has attracted great attention and interest from national defense departments and research institutes. Recently, the concept of time-domain channelization [15–17], which is based on the mapping relationship between time and wavelength by using dispersive devices, has been put forward. Photon channelization reception tends to be diversified. Various technologies emerge in an endless stream and continue to improve and update. High-performance microwave photonic channelized receivers depending on their large instantaneous receiving bandwidth and high-frequency resolution will have important applications in radar communications in the future.

1.2.2 *Challenges and Development Trends of Photonic Channelized Reception Technology*

(1) Development Trend of Photonic Channelized Reception Technology in the Future

The tremendous advantages of microwave photonics in the transmission and processing of broadband radio frequency signals breed photonic channelized reception technology, and the high attention and extensive research in various countries have made various channelized technologies more and more mature. Facing the signal environment with high density and complexity and the enlargement of frequency coverage, the photonic channelized receiver not only guarantees 100% interception probability but also has the ability of non-distortion processing to the changeable signal, which can accurately recover the original signal; while increasing the instantaneous reception bandwidth, it also needs to ensure high-frequency resolution and realize accurate frequency sensing. How to achieve high-precision reception of broadband and multi-standard radio frequency signals is a major research direction of channelized reception technology. Dynamic range is an important index to measure the receiver's signal processing ability. Photoelectric modulators, photoelectric converters, optical amplifiers, auxiliary RF devices and even optical fiber transmission media have nonlinear effects in the optical radio frequency link. Various nonlinear effects and the superposition of link noise have a great influence on the dynamic range and frequency resolution of the system. How to suppress all kinds of distortion and linearize analog optical links is another research direction to improve the performance of channelized receivers. With the rise and development of silicon-based optoelectronics, which integrates fusion photonic technology and electronic technology, the integration of various optoelectronic modules and optoelectronic systems has become a research hotspot. Silicon-based integration of channelized receivers provides the possibility of system integration, miniaturization and even implementation of on-chip systems. Therefore, making full use of the advantages of microwave photonic technology in broadband microwave signal transmission and processing, realizing a small integrated multifunctional microwave photon channelized receiver with broadband, large dynamic range and high-frequency resolution for radar, electronic warfare and communication applications will become the inevitable trend of radio frequency reception in the future.

(2) Challenges of Channelized Photonic Reception Technology

Currently, there are many schemes for ultra-wideband radio frequency photonic channelized reception, but they are faced with the challenge of difficult-to-achieve broadband separation, high accuracy and real-time coexistence. (1) The existing technologies are used to construct various fine filters in the optical domain. However, the huge difference between the THz and MHz radio frequency values results in the contradiction between the wide bandwidth of light and the accurate control of radio frequency. Optical filtering of MHz magnitude is difficult to achieve and its shape cannot be controlled, which makes it impossible to distinguish the signals with close carrier frequencies, or the crosstalk between channels is very large during channelization. On the other hand, frequency alignment between light sources and filters, between filters and filters, makes it difficult to accurately reach MHz magnitude, and either results in low accuracy of radio frequency sensing and large error or requires complex feedback frequency alignment. Therefore, in the channelized sensing of an optical radio frequency spectrum in UWB and multi-carrier, photoelectric fusion is faced with the problem of accurate spectral control. (2) Photonic channelized reception is oriented to broadband multi-carrier radio frequency signals. There are many kinds of devices in analog optical links with nonlinear distortion. Photoelectric fusion is faced with many kinds of non-linear conversion distortion coexisting, which greatly limits the dynamic range of links. 3) In order to reduce the sampling rate of subsequent ADC and the pressure of DSP processing, the broadband RF channel needs to down-convert each frequency band to lower IF. In UWB environment, down-conversion faces the problem of perception integration: on the one hand, integration can reduce device usage, simplify system scheme and reduce volume and power consumption; on the other hand, real time necessarily requires integrated perception and reception, such as separate perception and down-conversion design will inevitably spend more spectrum detection time, and channelized reception is confronted with the problems of separation of perception and reception, resulting in the redundancy of devices and poor real-time performance.

References

1. K. Chang, *RF and Microwave Wireless Systems* (Wiley, New York, 2000)
2. B. Razavi, *RF Microelectronics*, 2nd edn. (Pearson Education Inc., New Jersey, 2012)
3. X. Xie, Y. Dai, X. Kun et al., Broadband photonic RF channelization based on coherent optical frequency combs and I/Q demodulators. IEEE Photonics J. **4**(4), 1196–1202 (2012)
4. R. Li, H. Chen, Y. Ying et al., Multiple-frequency measurement based on serial photonic channelization using optical wavelength scanning. Optics Lett. **38**(22), 4781–4784 (2013)
5. H. Chen, R. Li, C. Lei et al., Photonics-assisted serial channelized radio-frequency measurement system with nyquist-bandwidth detection. IEEE Photonics J. **6**(6), 7903707 (2014)
6. R.E. Brooks, J.Z. Wilcox, SAW RF spectrum analyzer/channelizer using a focusing, phased array transducer, in *Proceedings of Ultrasonics Symposium* (IEEE, San Francisco, 1985), pp. 81–95
7. E.M. Alexander, R.W. Gammon, The Fabry-Perot etalon as an rf frequency channelizer, in *Proceedings of SPIE 0464, Solid-State Optical Control Devices* (SPIE, Los Angeles, 1984), pp. 45–52
8. J.P. Powers, J.P. Harvey, D. Marinsalta, Acousto-optic channelizer study, in *Proceedings of Ultrasonics Symposium* (IEEE, Honolulu, 1990), pp. 689–692
9. C.N. Ping, S. Dingrong, Acousto-optic heterodyne channelizer, in *International Conference on Communication Technology Proceedings* (IEEE, Beijing, 1996), pp. 571–574
10. W. Wang, R.L. Davis, T.J. Jung et al., Characterization of a coherent optical RF channelizer based on a diffraction grating. IEEE Trans. Microwave Theory Tech. **49**(10), 1996–2001 (2001)
11. S.T. Winnall, A.C. Lindsay, M.W. Austin et al., A microwave channelizer and spectroscope based on an integrated optical Bragg-grating Fabry-Perot and integrated hybrid Fresnel lens system. IEEE Trans. Microwave Theory Tech. **54**(2), 868–872 (2006)
12. S.J. Strutz, K.J. Williams, An 8–18-GHz all-optical microwave downconverter with channelization. IEEE Trans. Microwave Theory Tech. **49**(10), 1992–1995 (2001)
13. C.-S. Bres, S. Zlatanovic, A.O.J. Wiberg et al., Parametric photonic channelized RF receiver. IEEE Photonics Technol. Lett. **23**(6), 344–346 (2011)
14. Z. Li, H. Chi, X. Zhang et al., A reconfigurable photonic microwave channelized receiver based on an optical comb, in *International Topical Meeting on & Microwave Photonics Conference, Asia-Pacific (MWP/APMP)* (IEEE, Singapore, 2011), pp. 296–299

15. C. Wang, J. Yao, High-resolution microwave frequency measurement based on temporal channelization using a mode-locked laser, in *Microwave Symposium Digest (MTT)* (IEEE, Montreal, 2012), pp. 1–3
16. D.B. Hunter, L.G. Edvell, M.A. Englund, wideband microwave photonic channelised receiver, in *International Topical Meeting on Microwave Photonics* (IEEE, Seoul, 2005), pp. 249–252
17. J.M. Heaton, C.D. Watson, S.B. Jones et al., 16-channel (1- to 16-GHz) microwave spectrum analyzer device based on a phased array of GaAs/AlGaAs electro-optic waveguide delay lines, in *Optoelectronics and High-Power Lasers and Applications* (SPIE, San Jose, 1998), pp. 245–251

Chapter 2
Microwave Photonics

2.1 The Conceptual Connotation of Microwave Photonics

The microwave we refer to generally refers to electromagnetic waves with frequencies ranging from 300 to 300 GHz. Its frequency bandwidth is very large, so microwave communication has great potential. Because microwave band has many advantages, microwave band is used as communication band in ground mobile communication, military communication and aerospace communication system. The advantages of microwave communication system include:

(1) The bandwidth of the microwave band is large, and it has a bandwidth of nearly 300 GHz
(2) The base station uses the antenna to transmit signals, which has good structure and reconfigurability
(3) Transmission of microwave signals in the atmosphere, low transmission costs
(4) Receiver uses mobile devices to receive signals, which is very convenient.

Microwave communication also has some drawbacks which cannot be ignored: when the frequency increases, the loss caused by the long-distance transmission of microwave signals will be very large; the electromagnetic radiation intensity of high-frequency microwave signal is very large; and the microwave signal is unstable and vulnerable to environmental impact, which limits the development of microwave communication.

In the 1970s, photonics technology based on fiber-optic communication systems continued to mature. Different from traditional wave communication, optical fiber communication uses optical signals as carrier signals and transmits signals through the optical fiber. In fiber-optic communication systems, optical carrier signals are typically generated by lasers, and the amount of information they can carry is very large. As a transmission medium, optical fibers have many advantages, such as low transmission loss, wide bandwidth, anti-electromagnetic interference, small size, lightweight, low cost and so on. Although optical fiber communication has many advantages, its development is limited by its poor mobility.

J. Zhang and J. Li, *Satellite Photoelectric Sensing Technology*,
https://doi.org/10.1007/978-3-030-89843-4_2

Since the 1970s, with the continuous maturity and development of optoelectronic devices, optical fiber communication technology and microwave technology, a research field that combines the advantages of photon technology and microwave technology has gradually emerged. This field has developed and formed a new subject—microwave photonics. As a new subject, microwave photonics has a wide range of research fields, including optoelectronic devices, optically controlled microwave devices and systems working at microwave frequencies, photonic generation technology of microwave signals, terahertz spectroscopy technology, microwave photonic signal processing technology and microwave photon transmission links. Here, microwave refers to the microwave millimeter-wave band. Generally speaking, the above research fields of microwave photonics can be summarized into two aspects: one is to study optoelectronic devices and subsystems that can work at microwave frequencies; the other is to study how to apply these optoelectronic devices and subsystems to microwave photonic systems for microwave signal processing. The basic concept of microwave photonics is to use optical methods to process microwave signals. According to the definition of microwave photonics by Seeds and Willia Ills, it mainly includes two aspects:

(1) Study of photonic devices capable of processing microwave signals
(2) Application of optoelectronic devices and technologies to microwave systems.

Microwave photonics is an emerging interdisciplinary subject involving microwave engineering and photonics. It is designed to make full use of the advantages of photonics technology, such as low loss, lightweight, small size, broadband and anti-interference, to break through the electronic bottleneck encountered in microwave engineering so as to achieve functions previously impossible in analog or electrical domain. Because of these unique advantages, microwave photonics has become the focus of research and has been widely used in various fields. Figure 2.1 is a typical microwave photonic link or system structure [1].

At the input end, the analog microwave input signal is loaded on the optical carrier through electro-optic conversion, and the modulated optical carrier is transmitted to the receiving end through the optical transmission medium, and the microwave output signal is obtained after the photoelectric conversion at the receiving end. Compared with traditional microwave systems, microwave photonic systems have larger bandwidth, lower loss, lighter weight, and anti-electromagnetic interference, which make it possible to effectively solve the "electronic bottleneck" problem. Many functions that are difficult to achieve with traditional microwave systems can be realized with microwave photonic systems. Therefore, microwave photonics has attracted the attention of researchers in recent years and has been widely used in many fields such as national defense, communications, medical treatment and aerospace [2].

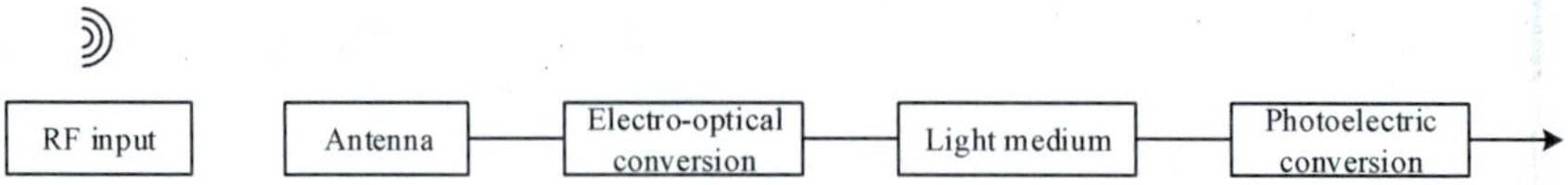

Fig. 2.1 Typical structure of microwave photonic links or systems

2.2 Advances in Microwave Photonics

Compared with coaxial cable, since its invention in the 1970s, optical fiber has become an important transmission medium in the field of communication, relying on its advantages of low loss (0.2 dB/km), anti-electromagnetic interference, small size and lightweight. Optical fiber communication technology using optical fiber as a transmission medium has made rapid development in the military and civil fields. With the continuous improvement of communication capacity and rate, optical fiber technology is attractive to the transmission and processing of radio frequency signals relying on its advantages of broadband and low loss. Either light wave or microwave is essentially electromagnetic wave. The consistency of theoretical basis promotes the integration of optical fiber technology and microwave technology, thus giving birth to a new interdisciplinary subject—microwave photonics [3]. Microwave technology can realize flexible, portable and low-cost wireless access. Optical fiber technology can be used for broadband radio frequency transmission with low loss and long distance. Combining the advantages of microwave photonics, the switching between light wave and microwave can be realized. Microwave photonics mainly takes millimeter wave as its research object, realizes the generation, transmission and processing of microwave signal by optical method, overcomes the electronic bottleneck of signal processing in a traditional electric domain, improves the sampling and control speed of the signal, and optimizes and improves the performance of microwave system in processing bandwidth, dynamic range, frequency response flatness and anti-electromagnetic interference ability. At the same time, the volume and weight of the system are reduced, and the cost of the system is reduced. With the rapid growth of signal reception bandwidth and transmission distance and the rapid expansion of information capacity, microwave photonics has attracted great interest and attention in the research and commercial field in many countries at home and abroad. The related application technologies and achievements emerge endlessly, and many functional devices have realized the commercialization of products.

As a subject dealing with high-speed photonic devices and optical/microwave systems for millimeter-wave processing, the research direction of microwave photonics is mainly divided into two aspects: on the one hand, high-speed optical devices applied in a millimeter-wave band, including photoelectric modulators, photodetectors, optical amplifiers and optical circuit integration and analog optical links; on the other hand, the study of microwave signal optics generation and control, including laser, photoelectric oscillator, the programmable optical processor of broadband microwave, microwave photonic filter, etc. [4]. In 1960, the world's first ruby laser with a wavelength of 694.3 nm was developed by Mayman of the Laboratory of Hughes Airlines, California. With the development of Erbium-doped gain fiber, the first mode-locked pulse laser with a repetition rate of 100 Hz was created by Southampton University in 1986 [5]. Many manufacturers in the world, including Coherent, IPG Photonics and TruFiber, have now commercialized the production of various high-performance lasers. In 1966, Gao Qin, the father of optical fibers, put forward the communication theory of using glass fibers as transmission media.

In 1970, Corning Company of the United States produced the first low-loss quartz optical fibers. Today, Ben Fujikura Company and China Changfei Company have introduced low-loss and low-dispersion optical fibers with losses below 0.18 dB/km. Since the 1950s and 1960s, the theory of electro-optic modulation has been studied. After decades of development, the fabrication technology of electro-optic modulator based on $LiNbO_3$ modulator, InP modulator and silicon modulator has become more and more mature. French Photline company and American EO Space company have commercialized the production of electro-optic modulator products with up to 60 GHz bandwidth. Since the advent of the photoelectric effect in 1873 and the development of semiconductor materials, photoelectric detectors with high responsiveness and bandwidth have emerged accordingly. In recent years, the worldwide communication upgrades, including the "Optical Reform" plan promoted by three major operators in China, the national high-speed optical fiber network project of Italy with 7 billion euros and the German "Digital Strategy 2025" layout, have brought revolutionary breakthroughs to the development of information field by using a light source as an information carrier and optical fiber as a transmission medium.

With the maturity and diversification of high-speed broadband optoelectronic devices, as well as the development of signal processing devices including optical wavelength division multiplexer, fiber Bragg grating, optical delay line and so on, the transmission and processing of radio frequency signals are more reliable. With the penetration of microwave photonics technology, people begin to devote themselves to the research of system level. The basic functional framework of the system structure is shown in Fig. 2.2, including optical mapping of microwave signals, radio frequency telescoping of low-loss optical fibers and digital reception after analog processing with broadband optical devices.

In the world, many companies including Emcore, Fnisar, Photonic System have developed RoF external modulation optical transceiver module with 20 GHz bandwidth. The Defense Department's Advanced Research Projects Agency (DARPA) has set up projects such as "Reconfigurable Microwave Photonic Signal Processor

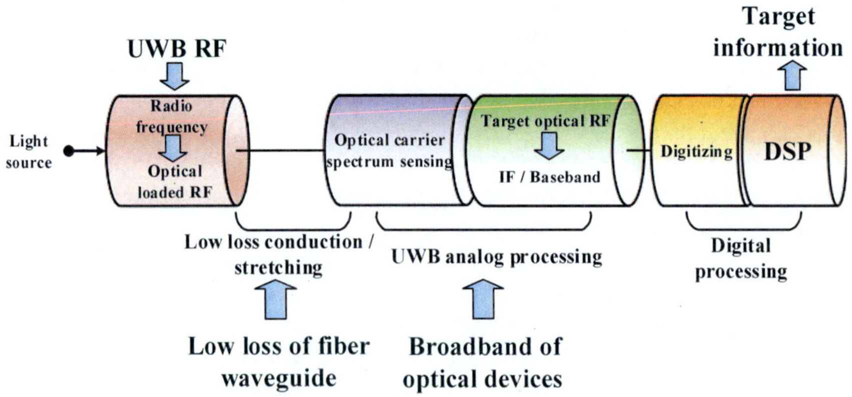

Fig. 2.2 System function module architecture based on microwave photonics

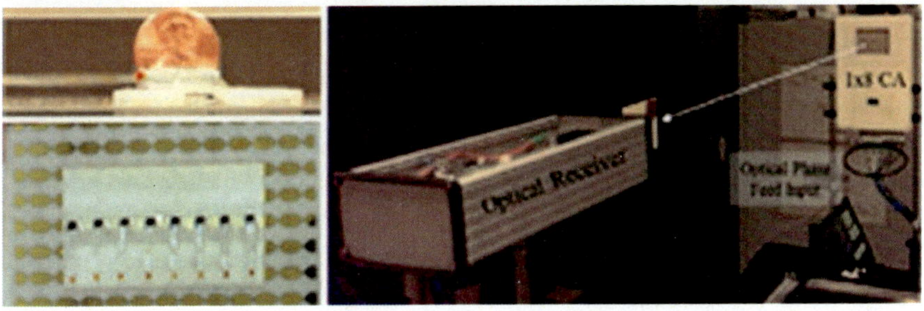

Fig. 2.3 1*8 Size of integrated array antenna and physical chart of integrated array transmitting/receiving antenna

(PHASER)" and "Photonic Bandwidth Compression Technology (PHOBIAC) in Large Instantaneous Bandwidth AD Conversion" to construct RF optical signal processing units. The University of California, Los Angeles, reported a photon time-stretched A/D system with 480 Gs/s and 96 GHz bandwidth. At present, the highest sampling rate is as high as 10 Ts/s. The Institute of Electronic Engineering of the University of Delaware, USA, has developed a 1 × 8 integrated photodiode-coupled array antenna with a 5–20 GHz ultra-wideband width and a positive and negative angle control range of 40°, as shown in Fig. 2.3 [6]. The TI Corporation completed advanced common aperture antenna arrays covering C ~ Ku band from 1990 to 1994 under the traction of the US Navy [7]. S.T. Winnall, a researcher at the Australian Defense Science and Technology Organization, has realized multi-carrier radio frequency spectrum analysis within 40 GHz bandwidth and with 90 MHz accuracy by scanning spectrum with optical filters. Researcher L.Nguyen has realized the identification of 20 and 40 GHz radio frequency points by constructing an optical Fourier transform [8]. Professor Capmany of Valencia Polytechnic University in Spain has carried out a comprehensive study of microwave photonics and has achieved landmark research results in simulating optical links in a large dynamic range. The European Union has also set up a series of major projects in the Seventh Framework Plan, such as "Low Cost Optical Access Network Technology and Optical Wireless Integration Technology (ISIS)". The optical processing project of microwave and digital signals carried out by ESA has realized satellite forwarding function by using microwave photon technology. The down-conversion from Ka band to C band, the generation of the adjustable local oscillator in the range of 26 GHz and the 4 × 4 non-blocking optical RF switching matrix forwarding have been realized successfully. The system-level ground demonstration has been realized [9]. In China, universities including Tsinghua University, Beijing University of Posts and Telecommunications, Zhejiang University, Nanjing University of Aeronautics and Astronautics, China Power Group and the aerospace sector have carried out in-depth research and achieved remarkable results in the field of communications. With the maturity of various technologies, microwave photonics will have broad application prospects in the field of national defense and military affairs.

2.3 Applications of Microwave Photonics

With the continuous innovation and development of various photonic technologies and the integration and commercialization of high-performance optical devices, microwave photonics has shown great potential in many fields such as civil communications, aerospace, microwave signal generation and processing. Following is a description of the key technologies that are now relatively mature:

(1) **Optical Wireless Fusion (RoF)**

With the rapid development of the Internet of Things and 4G mobile communication, many new wireless access services are pouring into people's work and life. Wireless communication technology has to increase the number of base stations and relay stations because of the serious signal attenuation when it propagates in the air, which leads to the problems of poor flexibility, high cost and difficult wiring. With the development of optical fiber communication, in order to support the existing broadband and ubiquitous communication services and meet the urgent requirement of the dramatic increase of data communication capacity, and at the same time to ensure the quality of signal transmission and the high utilization of network resources, the optical-borne wireless fusion technology, which combines optical fiber communication and wireless communication, emerges. Optical wireless transmission system uses optical fiber as the transmission medium to load data signals at the central station and transmits them through long-distance low-loss optical fiber links. In the remote antenna unit, the signals are radiated to the user cell through photoelectric conversion. The basic structure of the system is shown in Fig. 2.4. Compared with traditional wireless communication, it shortens the distance between users and antennas and saves at least 70% of power consumption. At the same time, the optoelectronic conversion between the central station and the base station can realize the

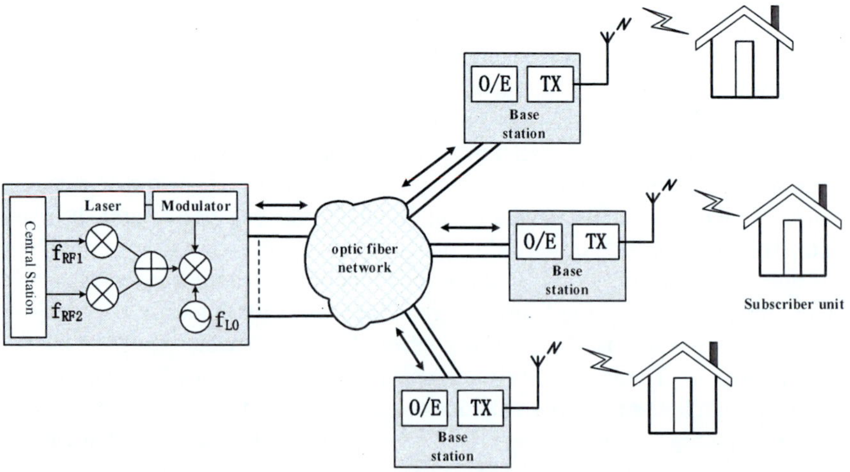

Fig. 2.4 Typical link of optical wireless fusion system

up/down-conversion function of the radio frequency signal in the optical domain, which simplifies the antenna structure and reduces the system cost [10]. The optical broadband also enlarges the data bandwidth of the communication and enhances the communication capacity. Optical wireless fusion technology concentrates signal processing in the central station, which makes system maintenance simple and easy to troubleshoot and centralize management. RoF technology, which relies on its advantages of broadband, low loss and high flexibility, has great potential in future broadband access networks such as indoor wireless coverage and next-generation wireless communication.

(2) **Stable-phase transmission technology based on optical fibers**

In the last 10 years, the accuracy of atomic frequency standard has improved rapidly, but the time and frequency transfer technology with high performance has developed slowly. Time and frequency transfer (synchronization) technology is an important support to improve the performance of many communication systems at present and in the future. It provides an important performance guarantee for many long-distance and distributed systems, such as space TT&C system and multistatic radar system. Traditional time–frequency synchronization schemes mainly include global positioning system (GPS), satellite bidirectional time transfer (TWSTT), wireless transmission, direct cable transmission and other schemes. The best performance of the GPS system and satellite bidirectional transmission system can only provide 10^{-15} stability, and the timing accuracy is about 1^{-10} ns. Comparatively speaking, the time–frequency phase stabilization technology based on optical fibers can avoid the influence of environmental temperature, humidity and other factors. It can achieve high-performance short-term frequency stability. Without any additional compensation, it can achieve 10^{-12} order of second stability. The corresponding control of the increase of optical fibers parameters can achieve nearly 10^{-18} order of sky stability, which is to achieve high stability frequency in the future. The preferred scheme of transmission and distribution. In 2001, the National Time Service Center of the Chinese Academy of Sciences took the lead in proposing the idea of precise time transmission using optical fibers. In 2009, the Time Service Center carried out an experiment of 100 km open-loop and simultaneous transmission of optical fiber time–frequency signals. In 2012, Tsinghua University and National Institute of Metrology (NIM) jointly carried out the simultaneous transmission of frequency and time channels on 80 km urban optical fibers, achieving 7×10^{-15} s stability and 5×10^{-19} days stability [11]. The time-stabilizing technology based on a tunable light source and optical fiber dispersion proposed by the Beijing University of Posts and Telecommunications realizes "passive high-precision on-orbit spacecraft line interferometry" on the baseline of 5.5 km in length. It has been applied to the "radio lunar data receiving and processing system" of Beijing Aerospace Flight Control Center. The X-band beacon signal of the Chang'e-3 satellite has been successfully collected. The measurement system is shown in Fig. 2.5.

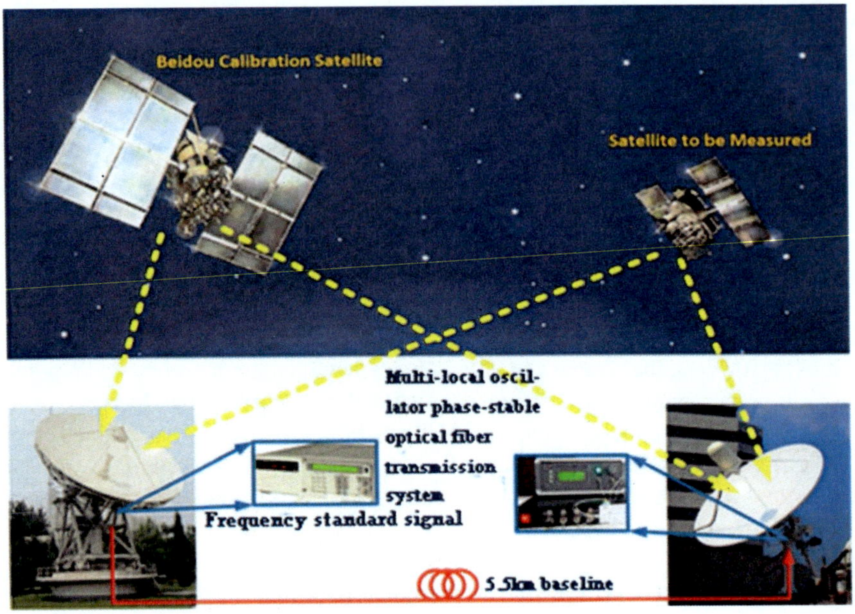

Fig. 2.5 A 5.5-km baseline interferometry system

(3) High-Performance Photogenic Microwave Technology

The broadband tunable microwave source with high-frequency spectrum purity has very important applications in a radar system, space navigation system, analog-to-digital conversion and accurate scientific measurement system. The stability of radio frequency signal is very important to the accuracy of radar ranging, the resolution of radar detection and the sensitivity of space TT&C. Traditional microwave oscillators use dielectric resonators and quartz crystals to generate radio frequency signals. The high-frequency characteristics and phase noise of the signals cannot meet the current application requirements. Microwave photon technology provides a new way of photomicrowave generation, including optical heterodyne method, external modulation method, four-wave mixing method, optical phase-locked loop method and photoelectric oscillator method [12]. It can not only avoid the processing technology of semiconductor materials and electronic devices but also be independent of the performance of various electronic devices. It can produce high performance of hundreds of GHz broadband tunable low phase noise of local oscillation signal. Optical modulation mainly uses the nonlinearity of the modulator to achieve the frequency doubling effect. The bandwidth of the modulator and the RF power amplifier limits the range of the generated RF signal. The principle of four-wave mixing is to use the nonlinearity of optical fiber medium to achieve high-order frequency doubling. However, this kind of scheme cannot avoid the influence of stimulated Brillouin scattering of nonlinear optical fibers, which limits the light efficiency, and produces large phase noise of signals. The phase-locked loop (PLL) method can improve the phase noise of the signal, and the power and frequency of the signal are easy to tune. The premise is that high bandwidth PLL and two narrow linewidth lasers are needed, which increases the cost and complexity of the system. Photoelectric oscillator (OEO) is a new type of oscillator, which converts light energy into microwave energy by a closed resonator and outputs it in the form of periodic signals. OEO is outstanding in the generation of microwave signals and has great application prospects in many fields [13]. OEO can produce microwave signals with high-frequency spectral purity by utilizing the low loss and broadband characteristics of optical fibers, and the noise of the signals is independent of the frequency. This characteristic is different from the traditional microwave signal based on the frequency doubling scheme. The phase noise of the high-frequency signal obtained by the frequency doubling scheme increases with the increase of the frequency spectrum. In order to obtain a high Q photoelectric cavity, a long optical fiber is usually needed as the energy storage medium. Due to the introduction of long optical fibers, the free spectrum range (FSR) of the cavity is small, and more longitudinal modes can be supported in the cavity. In order to make OEO oscillate in a specific frequency, it is usually necessary to introduce a filter in the cavity. By adjusting the center frequency of the filter, OEO with adjustable frequency can be obtained. Nowadays, OEO based on filters has been widely studied and exhibited greater tunability. Figure 2.6 shows the latest OEO product of OEwave Company in the United States. It has excellent performance. Its adjustable range is from 8 to 12 GHz, and its phase noise reaches −163 dBc/Hz at 10 kHz.

(4) **Optically Controlled Phased Array Radar Technology**

As a key means of identification, command, intelligence and control in military communications, phased array radar has rapidly occupied the radar market with its advantages of flexible beam scanning, high scanning data rate, high radiation power and high gain, and has become the mainstream application [14]. With the help of antenna technology, phased array radar has many antenna array units. Phase shifter or real-time delay line (TTD) is used to change the phase of the signal. Without rotating the antenna, it can control the direction of the beam without inertia by electric scanning. The response is flexible and the corresponding time is short. Different antenna array units can form different beams and achieve multi-function and high-precision acquisition. The data transmission rate is higher when multi-target information is

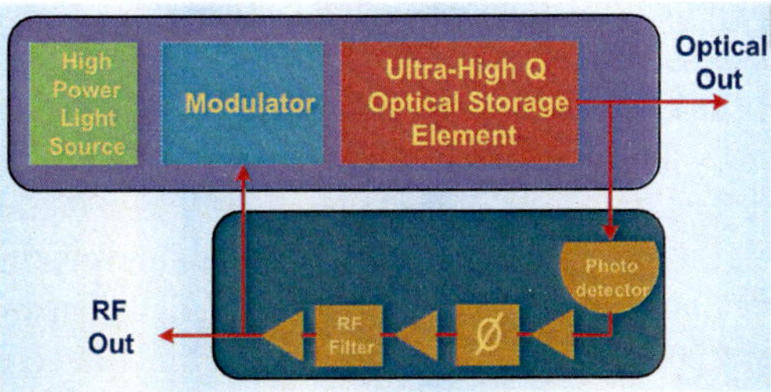

Parameters	Specification (Typ.)
Output Frequency	8 to 12 GHz
Phase Noise	- 75 dBc/Hz @ 10Hz - 115 dBc/Hz @ 100Hz - 145 dBc/Hz @ 1kHz - 163 dBc/Hz @ 10kHz -165 dBc/Hz @ 100kHz -170 dBc/Hz > 1MHz
Stability	10^{-11} @ 1 Sec.
	0.005 ppm @ 1 hr.
Package Size	19" Rack Mount

Fig. 2.6 AOEO structure principle and some performance parameters produced by OEwave Company, USA

taken. However, the traditional phased array radar based on phase shifter is affected by spatial and temporal dispersion in wide-angle scanning. The scanning range of the beam is small and the instantaneous bandwidth of the array is limited. Coaxial cable is used in delay line and other phased array devices. For large aperture and wide-angle scanning phased array antenna, long coaxial cable is needed. The signal power loss is serious, and the antenna structure is complex and the cost is high. With the development of microwave photonics, the application of optoelectronic technology in radar has produced the technology of optically controlled phased array radar, which provides an effective way to improve the overall performance of radar [15]. The optical phased array radar first modulates the radar signal to the optical carrier, then transmits it through the optical fiber and divides it into multiple channels with couplers. After each of them passes through different optical delay arrays, different phases are introduced. Then, the radar signal is recovered by photoelectric conversion and distributed to different antenna arrays by radio frequency amplification to radiate outward. Optical phase-controlled radar uses optical delay OTTD instead of coaxial cable, which reduces the power loss of signal, reduces the influence of transit time and aperture effect, thus realizes the radiation or reception of the large instantaneous broadband signal, and solves the bottleneck of wide-angle scanning [43]. The transmission of radio frequency signal by optical link is advantageous to the transmission and distribution of the signal, improves the stability of the signal, avoids the leakage of the signal and has an excellent concealment effect. It can extend the distance between the antenna and the signal processing center, realize the remote control of the radar and guarantee the safety of the radar and the staff. The substitution of coaxial cable eliminates the interference between antenna elements, enhances the anti-electromagnetic interference ability, simplifies the system structure and reduces the cost. With the development of optical signal processing, optoelectronic devices and optical integration, optical phased array radar will become an inevitable choice for a millimeter-wave radar system.

(5) **THz Time-Domain Spectroscopy(THz-TDS)**

The electromagnetic wave from 100 GHz to 10 THz band is called THz wave, which connects millimeter wave and infrared light. It is the transition region between electronics and optics. Because there is many chemical composition information in the spectrum, the research of THz technology has far-reaching significance for material exploration. With the development of nonlinear optics, materials science and microwave photonics, THz can be produced either optically or electronically. With the deepening of research, THz can be produced by photoconductive antenna, nonlinear crystal optical rectifier, air plasma, optical stimulated effect, optical frequency doubling and difference effect, quantum cascade laser and free electron laser. It can even reach 1 GHz spectral resolution. Terahertz can be divided into continuous terahertz and pulsed terahertz. Compared with microwave wavelength, narrow beam, good direction, heterodyne reception, THz communication is more secret and safe, and imaging resolution is high. Compared with infrared light [16], THz has strong penetration to non-polar substances, strong absorption and dispersion to most molecules, and can be used for material identification and perspective imaging.

Compared with X-ray, it has low radiation power. It does not cause photoionization of biological tissues and is suitable for medical imaging and security inspection systems.

Terahertz technology has been widely used in many fields since its rise in the 1980s after more than 10 years of research. In the field of public safety, most of the contrabands such as explosives, drugs and control tools have molecular fingerprint spectrum in terahertz band, which can be effectively identified and screened [59–63]. Terahertz has a strong penetrating force for non-polar substances such as clothing and dielectric materials. Coherent imaging can show the outline of hidden objects and obtain information on substance composition. Terahertz can be used for quality inspection of packaged goods and personal safety monitoring, providing an efficient means of public area security inspection. In food and drug detection, it can qualitatively identify different drugs, and at the same time, quantitative calculation of drug composition can be made to realize the detection and control of drug quality. Food safety is the focus of the public. Terahertz technology can detect and identify meat, encapsulated food and crop seeds. In the field of medicine and biochemistry, it can effectively identify different proteins, DNA, RNA and other macromolecular biological components for DNA detection. At the same time, it can also detect surface biological tissues, and carry out pathological identification and monitoring through medical imaging.

(6) Microwave Photon Signal Processing Technology

Microwave photon technology has great advantages in radio frequency signal modulation, frequency conversion, filtering, sampling and frequency measurement. The traditional up/down-conversion of signals is realized by the nonlinear characteristics of mixers such as field-effect transistors. Due to the factors such as bandwidth loss leakage interference of electronic circuit devices, the conversion efficiency in a high-frequency band is low and the signal noise is large [17]. Frequency conversion technology in the optical domain can modulate radio frequency signal to light by direct modulation, optical locking or heterodyne modulation. Any intermediate frequency conversion can be realized by optical heterodyne coherent detection. In addition, the frequency conversion of microwave signals can also be realized by using the nonlinear effect of optical devices, such as the cross-gain modulation of semiconductor optical amplifiers, the cross-absorption effect of electro-absorption modulators and the stimulated Brillouin scattering effect of gratings. With the increase of application frequency, microwave photonic filter emerges as the times require. The filtering of the microwave signal is transferred to optical frequency, which makes full use of the huge bandwidth of optical devices and enlarges the signal processing range. Photonic filters mainly use tunable optical delay lines or tunable laser sources and dispersion devices to introduce different delays. Compared with traditional filter devices, the tunable range of the filter is increased. Changing the output light power of the light source can change the tap coefficient of the filter, thus controlling the shape of the filter, making the filter reconfigurable and increasing the flexibility. Nowadays, a variety of band-pass, band-stop and two state-switchable microwave photonic filters with high Q values have been applied to microwave signal processing.

With the increase of communication bandwidth, the processing speed and capability of microwave signals are also limited by the back-end analog-to-digital conversion module. Optical-to-digital conversion provides a new idea for the digital processing of radio frequency signals. Optical sampling is provided by a mode-locked laser. The pulse width of the current pulse reaches femtosecond magnitude, GHz at repetition frequency, and pulse jitter is as low as hundreds of attoseconds. Compared with the traditional electric sampling pulse, it can provide a higher sampling rate, process wider bandwidth and achieve higher sampling accuracy [18]. Optical sampling mainly includes photoelectric hybrid sampling technology and all-optical sampling technology. Photoelectric hybrid analog-to-digital conversion (PADC) uses only optical pulse to sample and generate sampled optical sequences. After photoelectric conversion, electro-ADC completes quantization and coding. The high sampling rate is increased by high repetition frequency pulse, and high conversion accuracy is provided by low rate electro-quantization. Photoelectric combination improves the performance of the whole system. It is the most mature optical sampling technology currently studied. All-optical sampling is a process of extracting the amplitude and phase information of microwave signal from the sampled optical sequence by optical means, obtaining the variables corresponding to the signal, quantifying and coding the variables, thus realizing analog-to-digital conversion. All-optical sampling does not need the parameters of photoelectric conversion and ADC [19]. It directly derives digital information from the optical sampling sequence, avoids the errors introduced by photoelectric conversion and subsequent electrical processing, avoids the bottleneck of electronic circuits and greatly improves the accuracy of analog-to-digital conversion, and will become the mainstream direction of analog-to-digital conversion.

Photon-assisted instantaneous frequency measurement and broadband radio frequency signal reception technology have aroused widespread interest with the rapid increase of signal bandwidth and rate. Instantaneous frequency measurement technology can be divided into two categories according to the different core devices: frequency measurement technology based on dispersion medium (DE) and frequency measurement technology based on interference. DE scheme is based on the different group delay introduced by different wavelengths after modulating radio frequency signal passing through the dispersive devices to achieve spectral line separation so as to achieve microwave frequency measurement, but it does not have advantages in frequency measurement speed and system cost. Frequency measurement technology based on delay interferometry includes frequency-radio power mapping and frequency-optical power mapping. The power attenuation function is constructed by an interferometer to obtain frequency response. Signal frequency is determined according to power ratio, and high-speed frequency measurement is realized. However, frequency measurement technology based on power comparison can only measure a single frequency signal. Under multi-carrier input, it cannot be separated according to the output value. There is no sense in analyzing the input frequency. With the application of ultra-wideband and multi-carrier signals, broadband measurement technology with high sensing accuracy and real-time reception and processing has become the current development demand. The photonic channelized receiving technology, which combines the advantages of high precision of photoelectric technology and wide bandwidth of channelized reception, emerges at the historic moment. This technology physically separates the different frequency bands of the optical radio frequency spectrum and detects signal power through the photoelectric conversion matrix subchannel so as to obtain spectrum information. It has great advantages in receiving bandwidth and sensing accuracy in real time and will become the future radio frequency connection, the inexorable trend of receiver development.

2.4 Key Technologies of Microwave Photonics

Microwave photonic link is a kind of signal transmission technology that transmits radio frequency signal through optical fiber at transmitter and receiver. Its key is electro-optic conversion technology that converts radio frequency signals into optical signals and photoelectric conversion technology that converts an optical signal into an electrical signal.

2.4.1 Electro-optic Conversion

There are two main ways to load radio frequency signal into optical signal: direct modulation and indirect modulation. Direct modulation is to control the output optical power by using the RF signal as the driving voltage of the laser so as to load the RF signal into the optical signal. Due to the influence of chirp, continuous oscillation and electro-optic delay, direct modulation is seldom used in practice. Indirect modulation is the conversion of microwave signal to an optical signal by means of an external modulator, so it is also called external modulation. External modulation includes intensity modulation (IM) and phase modulation (PM). External modulation can eliminate the influence of chirp and oscillation. In view of its advantages and practicability, this section mainly analyzes the intensity modulation and phase modulation.

(1) **Intensity modulation**

Most people use electro-optic modulators to achieve intensity modulation. Among them, the most commonly used is the Mach–Zehnder interferometer intensity modulator (MZM) based on lithium niobate. As shown in Fig. 2.7, the MZM intensity modulator consists of a Y-splitter, two optical waveguide arms and a combiner.

Assuming that the optical signal of the input MZM is monochromatic light, it is expressed as

$$E_{in} = E_0 \exp(j\omega_c t). \tag{2.1}$$

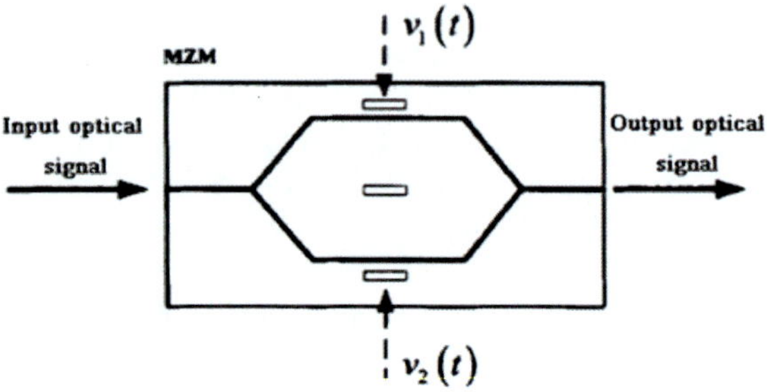

Fig. 2.7 Mach–Zehnder modulator

Among them, E_0 is the amplitude of input light field intensity and ω_c is the frequency of the optical carrier. External voltages (ω_{RF1} and ω_{RF2}: the angular frequencies of the upper and lower arm RF signals, V_1 and V_2 as amplitudes, φ_{RF1} and φ_{RF2} as initial phases) are applied as shown in formulas (2.2) and (2.3):

$$v_1(t) = V_1 \sin(\omega_{RF1}t + \varphi_{RF1}) \tag{2.2}$$

$$v_2(t) = V_2 \sin(\omega_{RF2}t + \varphi_{RF2}). \tag{2.3}$$

Optical phase changes such as those shown in formulas (2.4) and (2.5) can be generated in the upper and lower arms:

$$\phi_1(t) = \frac{\pi V_{DC1}}{V_{\pi 1}} + \frac{\pi V_1}{V_{\pi 1}} \sin(\omega_{RF1}t + \varphi_{RF1}) \tag{2.4}$$

$$\phi_2(t) = \frac{\pi V_{DC2}}{V_{\pi 2}} + \frac{\pi V_2}{V_{\pi 2}} \sin(\omega_{RF2}t + \varphi_{RF2}), \tag{2.5}$$

where V_{DC1} and V_{DC2} are DC bias voltage, $V_{\pi 1}$ and $V_{\pi 2}$ are half-wave voltage of upper and lower arms of MZM, respectively, that is, the bias voltage of MZM when the phase of optical carrier changes π.

The light signals of the upper and lower arms are coupled through the combiner, and the output light intensity is expressed as follows:

$$\begin{aligned} E_{out}(t) &= \frac{\sqrt{2}}{2} E_{in}(t) \left(\frac{\sqrt{2}}{2} e^{j\phi_1(t)} + \frac{\sqrt{2}}{2} e^{j\phi_2(t)} \right), . \\ &= E_{in}(t) \cos\left(\frac{\phi_1(t) - \phi_2(t)}{2} \right) \exp\left(j \frac{\phi_1(t) + \phi_2(t)}{2} \right) \end{aligned} \tag{2.6}$$

The output optical power obtained from formula (2.6) is

$$\begin{aligned} I_{out}(t) &= |E_{out}(t)|^2 = E_{in}^2(t)\cos^2\left(\frac{\phi_1(t)-\phi_2(t)}{2}\right). \\ &= I_{in}(t)\cos^2\left(\frac{\phi_1(t)-\phi_2(t)}{2}\right) \end{aligned} \tag{2.7}$$

The corresponding intensity transfer function (transfer function) is

$$\begin{aligned} T_{MZM} &= \frac{I_{out}}{I_{in}} = \cos^2\left(\frac{\phi_1(t)-\phi_2(t)}{2}\right). \\ &= \frac{1}{2}+\frac{1}{2}\cos(\phi_1(t)-\phi_2(t)) \end{aligned} \tag{2.8}$$

In order to achieve different modulation characteristics, it is generally necessary to load DC bias voltage on both arms of MZM. By changing the DC bias voltage, the refractive index of the upper and lower arm waveguides of the MZM modulator can be changed, and the fixed phase shift can be achieved by utilizing the voltage-controlled characteristics of lithium niobate. For convenience, the fixed phase shift caused by DC bias voltage is represented by φ_1 and φ_2. $\Delta\varphi = |\varphi_1 - \varphi_2|$ is used to represent the phase shift difference caused by DC bias voltage. Formula (2.8) evolves into

$$T_{MZM} = \frac{1}{2}\left(1+\cos\left(\Delta\varphi + \pi\frac{v_1(t)-v_2(t)}{V_\pi}\right)\right). \tag{2.9}$$

It can be seen from formula (2.9) that the intensity transfer function of MZM is related to phase shift and half-wave voltage. With the increase of half-wave voltage, the intensity transfer function of MZM presents a rising cosine function, and the modulation region of MZM is determined by $\varphi\Delta$. As shown in Fig. 2.8, the intensity transfer function has three typical offset points: orthogonal offset point,

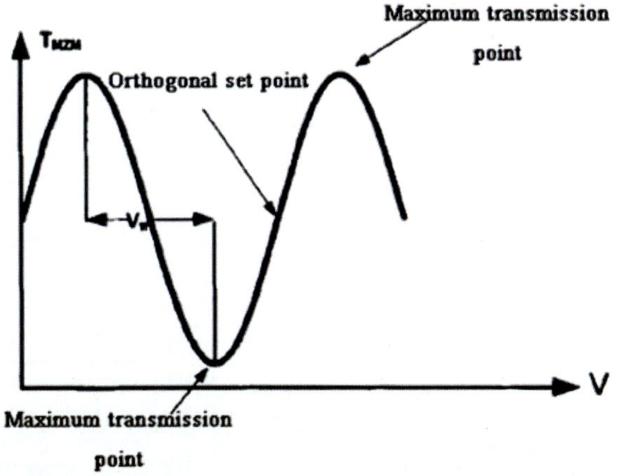

Fig. 2.8 Transfer function of MZM

maximum transmission point and minimum transmission point. MZM modulators exhibit different modulation characteristics in different modulation regions.

In practice, we usually use a single-arm MZM modulator. For a single-arm MZM modulator, the phase of the upper and lower arms can be expressed as

$$\phi_1(t) = 0 \tag{2.10}$$

$$\phi_2(t) = \frac{\pi V_{DC}}{V_\pi} + \frac{\pi V}{V_\pi}\cos(\omega_{RF}). \tag{2.11}$$

Among them, V_π is the half-wave voltage of single-arm MZM modulator, V_{DC} is the DC bias voltage, V and ω_{RF} are the amplitude and angular frequency of radio frequency signal, respectively, and the corresponding optical output signal is

$$\begin{aligned} E_{out}(t) &= \frac{1}{2}E_{in}(t) + \frac{1}{2}E_{in}(t)e^{j\pi\frac{V_{DC}+V\cos\omega_{RF}}{V_\pi}}. \\ &= \frac{1}{2}E_{in}(t)(1 + e^{j\pi\frac{V_{DC}+V\cos\omega_{RF}}{V_\pi}}) \end{aligned} \tag{2.12}$$

(1) Phase modulation

Compared with intensity modulation, phase modulation (PM) does not need to load bias voltage, which can avoid link instability caused by the disturbance of bias voltage. The basic structure of the phase modulator is shown in Fig. 2.9. By using the electro-optic effect of the crystal, the RF signal added to the arm can regulate the output light signal.

Assuming that the input PM monochromatic light is $E_{in}(t) = E_0(t)\exp(j\omega_c t)$ and the input microwave signal is $v(t) = V\cos(\omega_{RF})$, that is, the optical phase change is $\phi(t) = \pi V\cos(\omega_{RF})/V_\pi$, the output optical signal controlled by the microwave signal is as follows:

$$E_{out}(t) = E_0(t)\exp(j(\omega_c t + \phi(t))). \tag{2.13}$$

Usually, we define the modulation index as $m = \pi V/V_\pi$ and V_π as half-wave voltage of PM. According to Bessel expansion, we can simplify the output signal as follows:

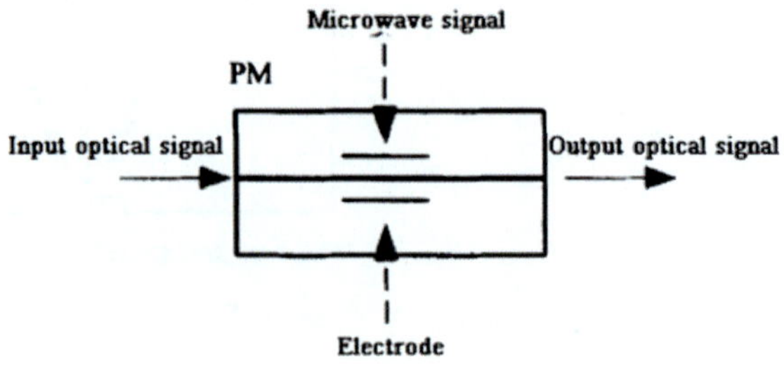

Fig. 2.9 Model of a phase modulator

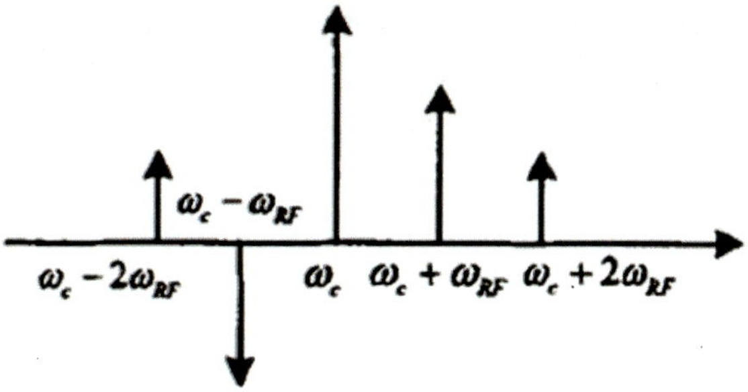

Fig. 2.10 Spectra after phase modulation

$$E_{out}(t) \propto J_0 e^{j\omega_c t} + J_1 e^{j(\omega_c t + \omega_{RF} t + \frac{\pi}{2})} - J_1 e^{j(\omega_c t - \omega_{RF} t - \frac{\pi}{2})}. \\ + J_2 e^{j(\omega_c t + 2\omega_{RF} t + \pi)} + J_2 e^{j(\omega_c t - 2\omega_{RF} t - \pi)} \tag{2.14}$$

Formula (2.14) shows that the spectrum generated by PM modulation also includes an optical carrier with frequency ω_c and harmonics with a frequency of integer multiple ω_m. The results are shown in Fig. 2.10. It can be seen that the harmonic amplitude remains unchanged and the phase difference between the positive and negative first-order sidebands is π.

2.4.1.1 Photoelectric conversion

(1) Operating Principle of Photoelectric Detector

Photodetector is one of the core devices of microwave photonic link. Its main function is to convert the optical signal of the link into an electrical signal. Suppose that the monochromatic light with angular frequency ω_c is

$$E(t) = E_0(t)\exp(j\omega_c t). \tag{2.15}$$

Among them, E_0 is the intensity of the light field, and the output current detected by the photodetector can be expressed as

$$i(t) \propto |E(t)|^2 = E(t)E^*(t). \tag{2.16}$$

$E^*(t)$ is the conjugate complex of $E(t)$. The photodetector will input the complex optical signal consisting of two optical signals in the form of (2.17) conjugate complex number and (2.18) conjugate complex number.

$$E(t) = E_1 \exp j(\omega_1 t + \phi_1) + E_2 \exp j(\omega_2 t + \phi_2) \tag{2.17}$$

$$E^*(t) = E_1 \exp j(-\omega_1 t - \phi_1) + E_2 \exp j(-\omega_2 t - \phi_2). \tag{2.18}$$

The output current measured by the photodetector is

$$\begin{aligned} i(t) &\propto |E(t)|^2 = E(t)E^*(t). \\ &= E_1^2 + E_2^2 + E_1E_2 \exp j(\omega_1 t - \omega_2 t + \phi_1 - \phi_2) \\ &\quad + E_1E_2 \exp j(\omega_2 t - \omega_1 t + \phi_2 - \phi_1) \\ &= E_1^2 + E_2^2 + 2E_1E_2 \cos(\omega_1 t - \omega_2 t + \phi_1 - \phi_2) \end{aligned} \tag{2.19}$$

Formula (2.19) shows that the output current consists of DC and AC terms, in which the magnitude of AC term current is $2E_1E_2$, the frequency is $\omega_1 - \omega_2$ and the phase is $\phi_1 - \phi_2$.

(2) Photodetector Detecting Intensity Modulation Signal

In the small-signal model, the optical output of microwave photonic link based on MZM intensity modulator can be approximately expressed as follows:

$$\begin{aligned} E_{out}(t) &\propto \exp(j\omega_c t) + J_0(m_{RF})\exp(j\omega_c t) \\ &\quad - J_1(m_{RF})\exp j(\omega_c t + \omega_{RF} t) - J_1(m_{RF})\exp j(\omega_c t - \omega_{RF} t). \\ &\quad + J_2 e^{j(\omega_c t + 2\omega_{RF} t + \pi)} - J_1 e^{j(\omega_c t - 2\omega_{RF} t - \pi)} \end{aligned} \tag{2.20}$$

The current detected by the photodetector is as follows:

$$\begin{aligned} i(t) &\propto |E_{out}(t)|^2 = E_{out}(t)E_{out}^*(t). \\ &= (1 + J_0)^2 + 2J_1^2 + 2_1^2\cos(2\omega_{RF} t) - 4(1 + J_0)\cos(\omega_{RF} t) \\ &= E_1^2 + E_2^2 + 2E_1E_2 \cos(\omega_1 t - \omega_2 t + \phi_1 - \phi_2) \end{aligned} \tag{2.21}$$

It can be seen from formula (2.21) that the output current includes DC term, ω_{RF}, $2\omega_{RF}$.

(3) Detection of Phase Modulation Signal by Photoelectric Detector

In the small-signal model, the optical output of the microwave photonic link based on the phase modulator can be approximately expressed as

$$\begin{aligned} E_{out}(t) &\propto J_0 e^{j\omega_c t} + J_1 e^{j\left(\omega_c t + \omega_{RF} t + \frac{\pi}{2}\right)} - J_1 e^{j\left(\omega_c t - \omega_{RF} t - \frac{\pi}{2}\right)}. \\ &\quad + J_2 e^{j(\omega_c t + 2\omega_{RF} t + \pi)} + J_2 e^{j(\omega_c t - 2\omega_{RF} t - \pi)} \end{aligned} \tag{2.22}$$

Therefore, the electrical signals detected by the photodetector are

$$\begin{aligned} I_{out}(t) &= |E_{out}(t)|^2 = E_{out}(t)E_{out}^*(t). \\ &\approx J_0^2 + 2J_1^2 + \left(J_1^2 - 2J_0J_2\right)\cos(2\omega_{RF} t) \\ &\approx J_0^2 + 2J_1^2 \end{aligned} \tag{2.23}$$

From formula (2.23), it can be seen that the electrical signal detected by the photodetector has only DC term, and does not include the ω_{RF} term, so it cannot restore the original radio frequency signal.

References

1. A.J. Seeds, K.J. Williams, Microwave photonics. J. Lightwave Technol. **24**(12), 4628–4641 (2006)
2. J. Capmany, D. Novak, Microwave photonics combines two worlds. Nat. Photonics **1**(6), 319–330 (2007)
3. J. Yao, Microwave photonics. J. Lightwave Technol. **27**(3), 314–335 (2009)
4. D. Marpaung, C. Roeloffzen, R. Heideman et al., Integrated microwave photonics. Laser Photonics Rev. **32**(16), 2849–2861 (2014)
5. I.P. Alcock, A.C. Tropper, A.I. Ferguson et al., Q-switched operation of a neodymium-doped monomode fibre laser. Electron. Lett. **22**(2), 84–85 (1986)
6. D.D. Ross, M.R. Konkol, C. Harrity et al., Integrated photodiode feeds for conformal UWB phased array antenna, in *IEEE International Topical Meeting on Microwave Photonics (MWP)* (IEEE, Long Beach, 2016), pp. 215–216
7. C. Hemmi, R.T. Dover, A. Vespa, Advanced shared aperture program (ASAP) array design, in *Phased Array Systems and Technology* (IEEE, Boston, 1996), pp. 278–282
8. S.T. Winnall, A.C. Lindsay, A Fabry-Perot scanning receiver for microwave signal processing. IEEE Trans. Microw. Theory Tech. **47**(7), 1385–1390 (1999)
9. M. Sotom, B. Benazet, A. Le Kemec et al., Microwave photonic technologies for flexible satellite telecom payloads, in *35th European Conference on Optical Communication* (IEEE, Vienna, 2009), 10.6.3.
10. A.S. Daryoush, E. Ackerman, R. Saedi et al., High-speed fiber-optic links for distribution of satellite traffic. IEEE Trans. Microwave Theory Tech. **38**(5), 510–517 (1990)
11. B. Wang, C. Gao, W.L. Chen et al., Precise and continuous time and frequency synchronisation at the 5×10^{-19} accuracy level. Sci. Rep. **2**(2), 556 (2012)
12. W. Li, J. Yao, A wideband frequency tunable optoelectronic oscillator incorporating a tunable microwave photonic filter based on phase-modulation to intensity-modulation conversion using a phase-shifted fiber Bragg grating. IEEE Trans. Microwave Theory Tech. **60**(6), 1735–1742 (2012)
13. D. Eliyahu, D. Seidel, L. Maleki, Phase noise of a high performance OEO and an ultra-low noise floor cross-correlation microwave photonic homodyne system, in *Proceedings of IEEE Conference on Frequency Control Symposium* (IEEE, Honolulu, 2008), pp. 811–814
14. R.A. Greenwald, K.B. Baker, R.A. Hutchins et al., An HF phased-array radar for studying small-scale structure in the high-latitude ionosphere. Radio Sci. **20**(1), 63–79 (1985)
15. B.-M. Jung, D.-H. Kim, I.-P. Jeon et al., Optical true time-delay beamformer based on microwave photonics for phased array radar, in *3rd International Asia-Pacific Conference on Synthetic Aperture Radar (APSAR)* (IEEE, Seoul, 2011), pp. 1–4
16. Y. Chen, C. Zhang, C. Hong et al., Optical frequency down-conversion from millimeter-wave to IF-band using an injection locked distributed feedback laser, in *OptoElectronics and Communications Conference* (IEEE, Vienna, 2009), pp. 1–2
17. K.-I. Kitayama, R.A. Griffin, Optical downconversion from millimeter-wave to IF-band over 50-km-long optical fiber link using an electroabsorption modulator. IEEE Photonics Technol. Lett. **11**(2), 287–289 (1999)
18. C. Schmidt-Langhorst, C. Schubert, C. Boerner et al., Optical sampling technologies and application, in *Optical Fiber Communication Conference* (IEEE, Anaheim, 2005), OTuG2

19. M. Sköld, M. Westlund, H. Sunnerud, All-Optical Waveform sampling in high-speed optical communication systems using advanced modulation formats. J. Lightwave Technol. **27**(16), 3662–3671 (2009)

Chapter 3
Spacecraft System

3.1 Classification of Spacecraft

A spacecraft is an aircraft that performs space missions such as exploration, development, or use of space in outer space outside the earth's atmosphere [1, 2].

According to whether manned or not, spacecraft can be divided into manned spacecraft and unmanned spacecraft, as shown in Fig. 3.1. Unmanned spacecraft mainly include artificial earth satellites, space (or deep space) detectors and cargo spacecraft. Manned spacecraft currently include manned spacecraft, space stations and manned transport vehicles [3, 4].

(1) Artificial earth satellite

Man-made earth satellites, or satellites for short, are unmanned spacecraft that orbit the earth. They circle the earth along a certain orbit under the action of the earth's gravitational field to make an unpowered flight. It is also the most launched spacecraft and the most versatile spacecraft. Satellites can have a variety of classification methods, including according to the characteristics of the orbit (such as altitude, orbit inclination and orbit precession), whether to recover, attitude control method, satellite weight etc. The commonly used ones are divided into scientific satellites, application satellites and engineering technology test satellites according to their uses.

Scientific satellites are used for scientific exploration and research, mainly including near-earth space physical exploration and astronomical satellites. Near-earth space physical exploration satellites are used to detect and study the upper atmosphere, ionosphere, earth's magnetosphere, earth's radiation belt, solar radiation, cosmic rays and aurora. Astronomical satellites are used to observe and study the sun and other celestial bodies.

Technology test satellites are satellites that carry out preliminary tests on some new technologies, new principles, new programs, new instruments and equipment that need to be applied to satellites. Some experiments that study the effects of

J. Zhang and J. Li, *Satellite Photoelectric Sensing Technology*,
https://doi.org/10.1007/978-3-030-89843-4_3

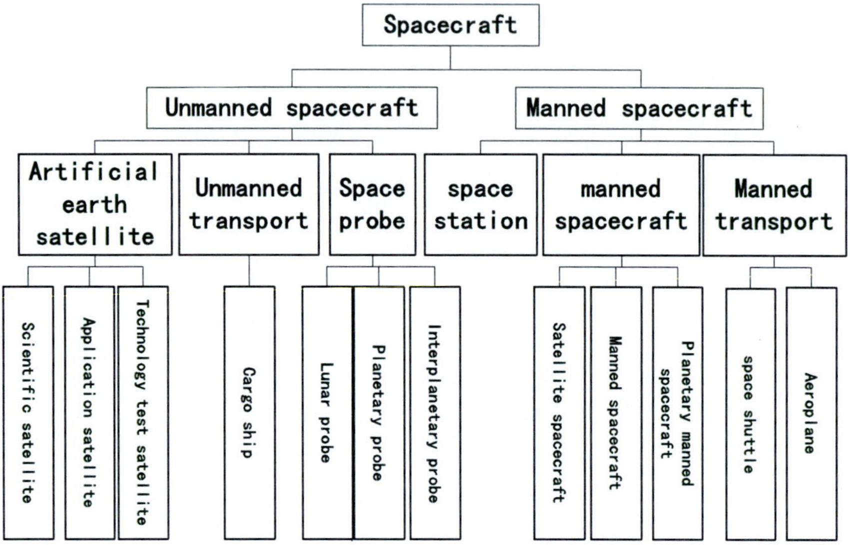

Fig. 3.1 Classification of spacecraft

microgravity and space environment on biology and manufacturing are carried out on manned spacecraft or space stations.

Application satellites are directly used for national economic or military purposes. According to its work characteristics and application capabilities, it is usually divided into remote sensing, communication and navigation satellites.

(2) Space probe

Space probes (also called deep space probes) are spacecraft that enter the gravitational range of other celestial bodies to detect other celestial bodies and space beyond the earth. Among them, those that land on the surface of other celestial bodies are called landers, those that enter the atmosphere of other celestial bodies for detection are called atmospheric detectors and those that orbit the celestial body are called orbiters, also called artificial satellites of the celestial body [5].

(3) Unmanned transport vehicle

At present, the main development is cargo spacecraft, that is, spacecraft used to carry cargo to orbit spacecraft or other celestial bodies.

(4) Manned spacecraft

Manned spacecraft currently include manned spacecraft, space stations and manned transport vehicles.

A manned spacecraft can ensure that astronauts live and work in space to perform space missions and return to the ground. According to the flight destination and orbiting celestial bodies, they can be divided into (earth) satellite-manned spacecraft, lunar payload spacecraft and planetary manned spacecraft. The space station is a large spacecraft in orbit that does not have the ability to return. It can be used by multiple astronauts to visit, work and live for a long time. The space shuttle is a type of manned spacecraft that has been out of service. It is a manned and reusable spacecraft that transports payloads between the surface of the earth and low-earth orbit. Currently, aerospace planes with more advanced functions and performances are being developed, which can repeatedly travel back and forth between the sky and the sky [6].

3.2 Main Application Satellites

(1) Communication satellite

The communication satellite transmits the radio communication signal through the transponder carried to realize the communication between the earth stations on the ground or between the earth station and the spacecraft. The emergence of communication satellites has brought about major changes in communication technology and has promoted the formation of new communication technology—satellite communication. Satellite communications have been widely used in international, domestic or regional communications, military communications, maritime communications, television broadcasting, and spacecraft tracking and data relay.

(2) Meteorological satellite

Meteorological satellites are artificial satellites for meteorological observation. Such satellites carry various meteorological remote sensors, which can receive and measure the visible light, infrared and microwave radiation of the earth and its atmosphere, and carry out a series of processing and analysis of the information to obtain various meteorological data.

(3) Earth resources satellite

Earth resources satellites use satellite-borne remote sensors to obtain electromagnetic wave information of various bands radiated and emitted by ground objects, and process and interpret this information to obtain the characteristics, distribution and status of various resources. Earth resources satellites play an important role in national land surveys, geological surveys, crop yield estimates, forest surveys, disaster monitoring, environmental protection, urban planning, and map surveying and mapping.

(4) Ocean satellite

Ocean satellites, also known as ocean observation satellites, are artificial satellites especially used for observing and researching the ocean. Such satellites carry marine remote sensors (mainly infrared and microwave remote sensors), which can receive electromagnetic wave information radiated by the ocean and can obtain various important information reflecting ocean phenomena and changes after processing.

(5) Navigation satellite

Navigation satellites are equipped with dedicated radio navigation equipment to directly provide navigation and positioning information such as precise position, speed and time to users on the ground, ocean, air and space.

The user receives the navigation and positioning information sent by the satellite and obtains the navigation parameters such as the user's distance relative to the satellite or the distance change rate through time ranging or Doppler velocity measurement. The users can determine their geographic location coordinates (two-dimensional or three-dimensional coordinates) and velocity vector according to the time and orbit parameters sent by the satellite to realize navigation and positioning.

Generally, a navigation satellite network (also called a navigation satellite constellation) is composed of multiple satellites, which can improve the three-dimensional coverage of global and near-earth space.

(6) Reconnaissance satellite

Reconnaissance satellites are artificial satellites used to obtain military intelligence. This type of satellite uses reconnaissance equipment such as photoelectric sensors or radio receivers to conduct reconnaissance, surveillance, or tracking of targets from orbit to collect intelligence on ground, ocean or air targets. The electromagnetic wave signal radiated, reflected or emitted by the target collected by the reconnaissance satellite is processed by image or signal processing, and various valuable information is extracted from it.

(1) Imaging reconnaissance satellite. Satellites are equipped with optical remote sensors or synthetic aperture radars to image and detect the target area from orbit. The information is processed and interpreted to obtain various military intelligence. In order to improve the resolution and the ability to identify targets, reconnaissance satellites generally operate in low orbits.
(2) Electronic reconnaissance satellite. Satellites are equipped with electronic equipment that collects and monitors radio signals, which are used to obtain signals emitted by the opponent's radar and telecommunications facilities. They also determine their geographic location or intercept their communications and provide relevant military intelligence.
(3) Missile early warning satellite. The satellite is equipped with a high-sensitivity infrared detection system. An early warning network is composed of several satellites, which can detect and track information about the launch of enemy missiles in time to realize early warning of missiles.

3.3 The Composition of the Spacecraft

A spacecraft generally consists of two parts: payload and platform. The payload is the core to directly complete the spacecraft mission. It is closely related to the application of spacecraft. Spacecraft for different purposes are equipped with different dedicated payloads, such as communication transponders and antennas of communication satellites, the navigation satellites of high-stability atomic clocks, radio beacons, various remote sensors for earth observation satellites (visible light cameras, infrared cameras, synthetic aperture radars, microwave radiometers etc.) and so on.

The spacecraft platform is a support system for spacecraft and payload work. Its main function is to provide corresponding structure, thermal control, orbit and attitude control, power supply, telemetry and remote control, and data management and transmission guarantee capabilities to support the payload to complete designated tasks in space according to user requirements.

At the same time, a spacecraft is an aircraft that is launched from the ground into space by a carrier, which orbits in space and works in a space environment. Therefore, the spacecraft platform is as small in size and light in mass as possible and can adapt to the environment such as vibration, shock, noise, overload, air pressure change during the launch. It is able to adapt to harsh space environments such as space electromagnetic radiation, thermal radiation, high-energy particles and vacuum to provide a working environment suitable for the normal work of the load.

Although the complexity of various spacecraft platforms is different, spacecraft platforms generally include subsystems such as structure and mechanism, thermal control, power supply, attitude and orbit control, measurement and control, data management etc. Some spacecraft also have propulsion, data transmission and return.

The function of the structural subsystem is to carry the instruments and equipment of each subsystem, to provide installation interfaces and protection for other subsystems and facilities, to ensure that the spacecraft has sufficient strength and rigidity throughout its life cycle, and to provide an interface with the launch vehicle. It generally includes the main bearing structure, auxiliary structure and connectors. The mechanism is a device that realizes and controls the movement of the movable components on the spacecraft, such as release mechanism, deployment mechanism, locking mechanism, driving mechanism, pointing mechanism, separation mechanism, derotation mechanism etc.

The task of the thermal control subsystem is to ensure the operating temperature range of each subsystem in each stage of the orbit operation. The technology of the thermal control subsystem can be generally divided into passive, semi-passive and active. The passive type refers to the method without moving parts or heaters, such as the use of multi-layer insulation materials, thermal control coatings, heat pipes and expandable heat radiators, etc.; the semi-passive type refers to the use of a simple control device driven by a thermal device or a motor-driven control device to open or close the heat conduction channel to dissipate heat, such as blinds; the active type refers to electric heaters, mechanical circulation pumps and in the refrigerator and so on.

The power subsystem is responsible for providing power to the electrical loads on the satellite during the working life cycle of the satellite; generally, it consists of a primary power subsystem and an overall circuit subsystem. In terms of the current domestic spacecraft power subsystem, it is also called the power supply and distribution system. The power subsystem must have functions such as power generation, energy storage, distribution, bus voltage regulation and battery charge and discharge control, and some also require secondary power sources with multiple voltages to be transformed and stabilized. At present, spacecraft widely use solar arrays and batteries called primary power supplies. In addition to the solar cell array and battery, the power subsystem also has power control equipment including battery charge and discharge controllers, solar cell array shunt regulators, bus regulation etc.

The attitude and orbit control subsystem is used to maintain or change the attitude and orbit of the spacecraft. Among them, the attitude control is to control the rotation of the satellite as a whole around its center of mass or the rotation of the components on the satellite relative to the satellite so that the relevant components of the satellite (such as solar wing, antenna, camera etc.) can obtain and maintain the orientation in space; orbit control is to control the satellite in orbit so that its center of mass enters and remains in a predetermined orbit [7, 8].

The attitude and orbit control subsystem generally consists of sensors (such as infrared, sun and star sensors, inertial sensors, magnetometers etc.), controllers (commonly used computers) and executive components (momentum wheel, flywheel, torque gyroscope, magnetic torque device, jet or dual-component engine etc.).

The data management subsystem is the information center of the spacecraft, which can realize the functions of satellite telemetry, remote control, orbit measurement and positioning with the cooperation of other subsystems and the ground. The concept of telemetry refers to the transmission of various measurement information on the satellite to the ground station by radio. Remote control refers to the transmission of related control instructions, data or programs from the ground to the satellite; orbit measurement and positioning can be combined the satellite's angle measurement, speed measurement and distance measurement through the ground station radio and satellite autonomous positioning and orbit determination based on the navigation constellation. Early spacecraft had independent telemetry subsystems, remote control subsystems, and orbit measurement and positioning subsystems. With the development of computer technology, the satellites continue to expand autonomous management and other functions, thus forming a unified on-board data management system, sometimes called the satellite service subsystem [9, 10].

There are also some subsystems specially set up by specific spacecraft to complete missions, such as return and recovery subsystems that are unique for returnable spacecraft. Environmental control and life support subsystems are necessary for manned spacecraft. The emergency response subsystem is used in emergency situations to enable astronauts to quickly leave the danger zone and return to the ground in time and safely.

References

1. SIA, State of the Satellite Industry Report. SIA: 2017(20)
2. R. Blockley, W. Shyy, *Encyclopedia of Aerospace Engineering*. Wiley (2010)

3. S.L. Padula, J.J. Korte, H.J. Dunn, A.O. Salas, Multidisciplinary optimization branch experience using iSIGHT software. NASA/TM-1999–209714 (1999)
4. D. Monell, D. Mathias, J. Reuther, et a1., Multi-discipli-nary analysis for future launch systems using NASA's Advanced Engineering Environment (AEE), in *AIAA* (2003), p. 3428
5. J. Dong, P. Motaghedi, D. Ngo et a1., Design to cost: an integrated optimal approach to reducing the number of parts required for the Boeing Delta IV tail service mast system, in *8th AIAA/USAF/NASA/ISSMO Symposium on Multidisciplinary Analysis and optimization* (AIAA, Long Beach, CA, September 6–8, 2000), p. 4915
6. A. Carty, An approach to multidisciplinary design, analysis and optimization for rapid conceptual design, in *AIAA* (2002), p. 5438
7. D.W. Miller, L.W. Alivier, E.M. Gary, Framework for multidisciplinary integrated modelling and analysis of space telescopes. Integr. Model. Telescopes (2002)
8. F. Xu, H.M,F. Tao, S.J. Cok, J. Cox Simon, Workflow support for advanced grid-enabled computing. [2005–9–8]. www.allhands.org.uk/2004/proceedings/papers/171.pdf
9. J. Rogers, A. Salas, R. Weston, A web-based monitoring system for multidisciplinary design projects, in *AIAA98* (1998), p. 4706
10. Phoenix integration accelerating product development through grid computing. CenterLink white Papar. [2005–9–8]. http://www.phoenix-int.com/

Chapter 4
Communication Satellite Technology

4.1 Introduction

The overall technology of communication satellite refers to the engineering technology for the overall design, analysis and verification of communication satellites by applying spacecraft system engineering theory and methods according to user mission requirements and engineering large-scale system interface constraints, including satellite mission demand analysis technology, orbit and constellation design technology, engineering large-scale system interface design technology, communication payload system design technology, satellite platform design technology, satellite overall performance analysis and optimization technology, system-level test verification technology, space environment protection design technology, reliability and safety technology, development process and standard system, etc. Its fundamental goal is to develop communication satellites that meet user mission requirements, coordinate and match with engineering large-scale system interfaces, technical implementation methods are reasonable and feasible, system comprehensive performance is optimal, and key functional performance can be tested and verified on the basis of existing technology [1].

4.2 Satellite Communication Services and Their Spectrum Allocation

1. **Characteristics of satellite communication services**

Satellite communication is the use of artificial earth satellites as relay stations to relay or reflect radio signals and to communicate between satellite communication earth stations (referred to as earth stations) or between earth stations and spacecraft. Satellite communication is an important result of the combination of aerospace technology and modern communication technology. It has been widely used in radio

J. Zhang and J. Li, *Satellite Photoelectric Sensing Technology*,
https://doi.org/10.1007/978-3-030-89843-4_4

and television, mobile communications and broadband internet, and is one of the essential communication methods of today [2].

The satellite communication system generally consists of a space segment, a ground segment and a control segment. Among them, the space segment mainly includes artificial earth satellites that act as radio relay stations in space orbits. Satellites that undertake this service are generally referred to as communication satellites. The ground segment mainly refers to the earth station, and there are various forms such as fixed, vehicle, shipborne and airborne according to the use mode. The control section contains the tracking, telemetry and remote control facility systems necessary for system operation. Figure 4.1 shows a schematic diagram of a satellite communication system. The space segment is a communication satellite. The earth station includes an air mobile station, a land mobile station and an ocean shipping station. The control section is configured by a ground core such as a central station and a communication hub composition [3].

Based on the provisions of the ITU Radio Regulations, several services often involved in satellite communications include the fixed-satellite service (FSS), the

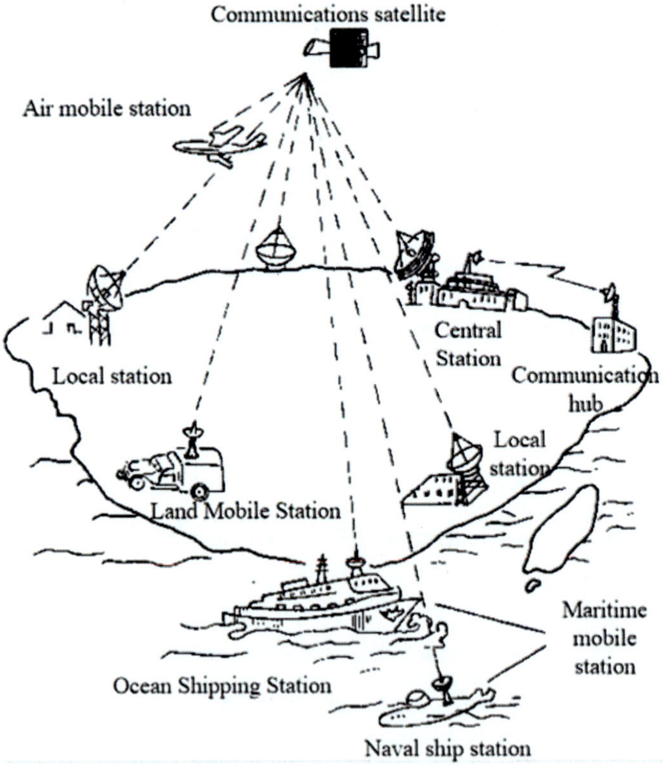

Fig. 4.1 Schematic diagram of satellite communication system

mobile-satellite service (MSS), the broadcast-satellite service (BSS) and the inter-satellite service (ISS) [4].

The fixed-satellite service is a communication service carried out between earth stations at a given location by using one or more satellites. The given location may be a designated fixed location or any fixed location within a designated area. The fixed-satellite service usually uses the C, Ku or Ka bands.

The mobile-satellite service is a communication service carried out between a mobile earth station and one or more satellites, or between mobile earth stations using one or more satellites. According to the different types of mobile earth stations, in practical applications, it can be considered that satellite mobile services also include satellite land mobile services, satellite maritime mobile services and satellite aeronautical mobile services. The mobile-satellite service usually uses the L or S band.

The satellite broadcasting service is a communication service that uses satellites to send or forward signals for direct public reception (including individual reception and collective reception). The satellite broadcasting service usually uses the C or Ku band.

The inter-satellite service is a communication service that uses satellites between multiple user spacecraft and is mainly used to forward the tracking and control signals from the earth station to the user spacecraft and relay the information sent back to the ground by the user spacecraft. The inter-satellite services usually use the S, Ka or Q/V bands [5].

Satellite communications can use geostationary orbit (GEO) satellites, medium-earth orbit (MEO) satellites or low-earth orbit (LEO) satellites as space segments. At present, there are a large number of communication satellites using geostationary orbit, but with the development of mobile and broadband communication services facing the world (including the south and north pole regions), satellite constellations in low-earth orbit are also vigorously developing.

The geostationary orbit is an orbiting, circular satellite orbit with an operating period equal to the earth's rotation period, an inclination of 0°, and an orbital height of 35,786 km. Low-earth orbit is a satellite orbit located hundreds of kilometers to 2,000 km above the surface of the earth. The medium-earth orbit satellite is a circular or elliptical satellite orbit with an altitude of 2000–35,786 km.

2. **Satellite communication service band**

Spectrum allocation for the satellite service is a fairly complex process that requires coordination and planning internationally. The spectrum allocation for the satellite service is carried out under the management of the International Telecommunication Union (ITU). To make frequency planning easier to implement, the ITU divides the entire earth into three regions, as shown in Fig. 4.1. Region 1: Includes Europe, Africa, the former Soviet Union and Mongolia; Region 2: Includes North and South America and Greenland; Region 3: Includes Asia, Australia and Southwest Pacific. Different frequency bands may be used in these areas. Some of the services provided by satellites can be divided into satellite fixed services, satellite broadcasting services, satellite mobile services, satellite navigation services and satellite weather services.

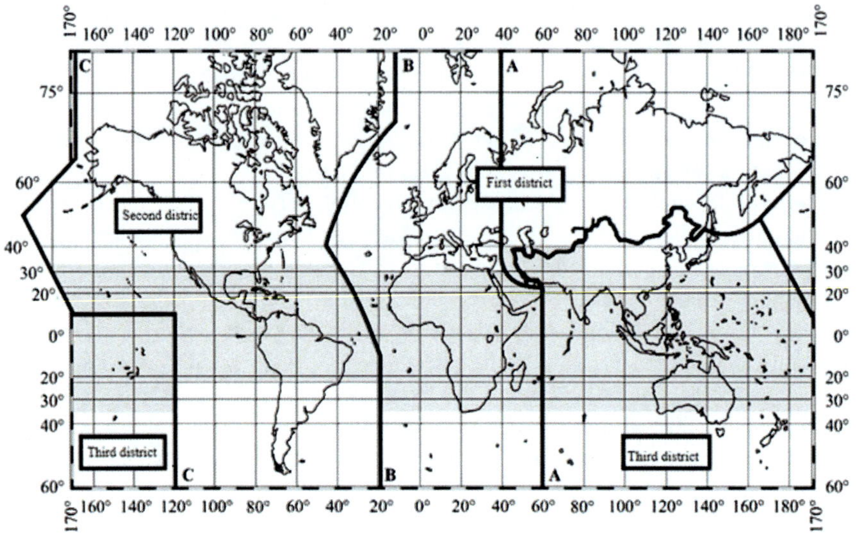

Fig. 4.2 Regional division of the International Telecommunication Union

In the above service classification, more detailed divisions can also be made. For example, the fixed-satellite service can provide a satellite link to an existing telephone network, or it can provide a satellite link to a cable television network company for transmitting television signals. The satellite broadcasting service mainly refers to live broadcast to the home, sometimes called the direct broadcast satellite (DBS) service. In Europe, it is also known as the direct-to-home (DTH) service. The mobile-satellite service may include land mobile, maritime mobile and aeronautical mobile. Satellite navigation services include global positioning systems, and satellites used in meteorological services are often used to provide search and rescue services [6]. Regional division of the International Telecommunication Union is shown in Fig. 4.2.

Table 4.1 lists the commonly used frequency bands for the satellite service. The Ku frequency band indicates a frequency band lower than the K frequency band, and the Ka frequency band indicates a frequency band higher than the K frequency

Table 4.1 Common service frequency bands for satellite communications

Frequency band	Frequency range (GHz)	Frequency band	Frequency range (GHz)
VHF	0.1–0.3	UHF	0.3–1.0
L	1.0–2.0	S	2.0–4.0
C	4.0–8.0	X	8.0–12.0
Ku	12.0–18.0	K	18.0–27.0
Ka	27.0–40.0	V	40.0–75
W	75–110	mm	110–300

band. The Ku band is currently used for the direct broadcast satellite service, and it is also used for some fixed services. The C band is used for the fixed-satellite service, and it is not allowed to use this band for live broadcast services. The VHF band is used for certain mobile and navigation services and is used for data transmission by meteorological satellites. The L band is used in the mobile-satellite service and the navigation service. For the fixed-satellite service in the C band, the most commonly used sub-band is 4–6 GHz. The higher frequencies are almost always used for land-to-satellite uplinks. The frequency range commonly used for live broadcast services in the Ku band is 12–14 GHz. Although the frequency division is very precise, in actual systems, it sometimes exceeds the frequency range given above. For example, the frequency used in the Ku band may be 11,730–14,030 MHz.

4.3 Communication Satellite Orbit and Constellation Design

The orbit is closely related to the mission of the satellite. According to the mission requirements of the satellite, selecting the most favorable orbit is the primary task of orbit design. The satellite's coverage of the earth, orbit life, the sun incidence law, eclipse and other factors closely related to the overall design all affect the selection of the orbit. According to different orbital height, the commonly used orbits of communication satellites are mainly divided into three categories: geostationary earth orbit (GEO), low-earth orbit (LEO) and medium-earth orbit (MEO), as shown in Fig. 4.3.

A satellite communication system can be composed of several communication satellites to form a communication satellite constellation. The constellation can include GEO, MEO and LEO communication satellites. Through the optimization

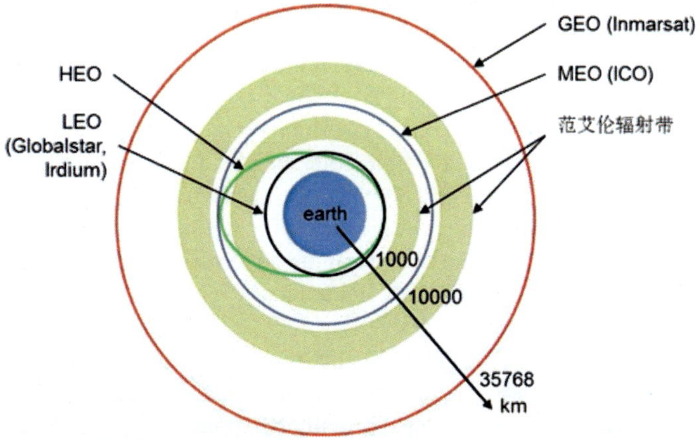

Fig. 4.3 Schematic diagram of satellite orbit types

of the constellation configuration, the communication system's wide coverage, high availability, enhanced regional coverage and system cost reduction can be achieved.

The orbit of a geostationary satellite is located on the earth's equatorial plane, with a height of 35,786 km, and an operating period of 24 h, which is the same speed as the earth's rotation. Therefore, geostationary orbit satellites are like stationary in the sky, which is very suitable for communication missions. GEO satellites have a large orbital coverage area. One GEO communication satellite can cover about 40% of the earth's surface. Three GEO communication satellites at equal intervals on the equator can achieve global communications except for the two poles. The coverage area of a GEO satellite is approximately equivalent to 10 LEO satellites. Major military, civilian or commercial communication systems in the world use GEO as their orbits, such as the US military broadband communication systems (DSCS, WGS), narrowband communication systems (UFO, MUOS), protection systems (Milstar, AEHF), direct broadcast satellites (Direct TV), data relay satellite system (TDRSS), maritime satellite system (INMARSAT), Asia Pacific satellite mobile communication system (APMT), my country's Zhongxing series communication satellites etc. [7].

LEO satellites have an orbit height of 500–2000 km and are located below the Van Allen belt. The operating cycle of the satellite is about 100 min, and the satellite viewing time is about 15 min. The orbit form can be polar or inclined orbit. The use of LEO as the orbit for communication satellites can reduce the power attenuation of communication links, reduce communication delays and simplify the design of satellites and user terminals. However, due to the low orbital height, the area covered by a single satellite is limited. Communication missions often need to be realized through satellite networking, with dozens of networking satellites. For example, the Iridium satellite system selects a near-circular orbit with a height of about 780 km, an orbital inclination of 86.4° and a total of 66 satellites for networking. There are also LEO communication satellite systems: the GLOBALSTAR system, which contains 48 satellites, with an orbital height of 1410 km and an inclination angle of 52°; the ORBCOMM system, which contains 32 satellites with an orbital height of 810 km, and the orbital inclination angle is divided into 45° and 0°. Figure 4.4 shows the coverage of the LEO communication satellite on the earth's surface.

The orbit of the medium orbit satellite is between the two Van Allen belts, and the height is between 8000 and 12,000 km. The orbital period is 5–12 h, and the satellite's visible time is 2–4 h. Because of the high orbital height, it will put a lot of pressure on the power amplifier on the satellite. Therefore, the medium orbit can take into account the advantages of low-orbit and geostationary orbit satellites. Typical medium orbit satellite constellations include the O3B communication satellite system of the United States, the United States GPS, China's Beidou and Russia's GLONASS navigation satellite system.

Three GEO satellites are sufficient to cover the entire earth except for the north and south poles. GEO satellites are very suitable for providing broadcast/multicast services with global coverage, as well as providing regional mobile and fixed services. MEO satellites and LEO satellites are non-stationary to users on the earth. Therefore, multiple satellites are needed to alternately provide coverage for a certain area on

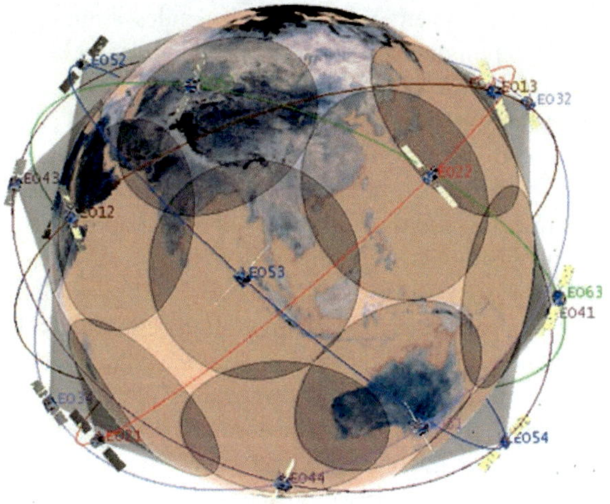

Fig. 4.4 Schematic diagram of LEO satellite ground coverage

the earth. A global MEO system requires 10–12 satellites to form a constellation to ensure that the minimum elevation angle is not less than 30°. The LEO system usually requires more than 40 satellites to ensure the coverage requirements of the minimum elevation angle of 10–40°. For MEO and LEO systems, the recommended minimum elevation angle is 40° so that higher link availability and acceptable delay changes can be obtained. Compared with GEO satellite systems, LEO and MEO satellite systems can obtain a smaller end-to-end data transmission delay. Compared with GEO satellites, MEO and LEO satellites generally have a smaller launch weight and relatively small size, which can achieve multiple satellite launches with one arrow, thereby reducing the cost of constellation construction and shortening the constellation network cycle.

4.4 Communication Satellite Payload System Design

Communication satellites can be divided into communication broadcasting satellites, mobile communication satellites, broadband communication satellites and laser communication satellites according to their business types. Among them, the communication and broadcast payload can provide services such as audio and video broadcasting directly to the home or even directly to the handset; the mobile communication payload can provide services such as voice, short message, data communication between personal handheld terminals; broadband communication payload can provide users with multimedia communication services such as voice, video, data, and images; laser communication payloads use the laser as an information carrier for communication, mainly providing high-speed data transmission services, which is an important development direction for communication satellites. The communication satellite payload mainly includes antenna and transponder subsystems; for laser communication satellites, it also includes aiming, capturing and tracking subsystems. Communication satellites also include tracking and data relay satellites, which mainly provide tracking and high-speed data transmission services from ground stations to satellites, spacecraft, aircraft etc., and can relay data and telemetry information provided by user spacecraft to ground stations, and can transmit information such as control instructions and data from the ground station to the user's spacecraft. The tracking and data relay satellite payload mainly includes antenna, transponder and acquisition and tracking subsystem [8, 9].

The design of the communication satellite payload system refers to the preliminary determination of the configuration of the payload subsystems under the specified frequency band, orbital position, service type, system capacity, user service area and other requirements, considering factors such as satellite life and reliability, etc. The satellite platform is selected according to the platform interface indicators such as the power and weight of the payload; according to the configuration characteristics of the satellite platform and various constraints such as the existing technical conditions, the payload configuration and configuration layout design, including the configuration of the antenna subsystem type and transponder subsystem configuration and communication cabin layout; according to the preliminary plan of transponder and antenna subsystem, optimize the allocation of system indicators. The above work content generally requires multiple iterations to complete the payload design.

(1) EIRP index decomposition

EIRP index decomposition refers to the determination of the gain of the transmitting antenna, the output power of the transponder power amplifier and the loss of the transponder output end. The antenna gain is basically determined by the antenna coverage requirements, but the design needs to consider the influence of antenna installation errors, structural thermal deformation and satellite attitude control performance. After the gain index of the antenna is determined, the output power index of the transponder is also determined accordingly. In the actual design, the interface between the antenna subsystem and the repeater subsystem should be clarified to ensure that the antenna gain and the output power index allocation of the repeater are not inconsistent. When the system margin in the preliminary design is insufficient, the detailed system design should minimize the feeder loss by optimizing the layout.

(2) G/T index decomposition

G/T index decomposition refers to determining indexes such as receiving antenna gain and receiving system noise temperature. The direction error and interface relationship should be considered when determining the gain index of the receiving antenna. In terms of noise temperature, the noise temperature of the receiver is limited by the current technical level, while the antenna noise temperature is mainly limited by the thermal radiation characteristics of the ground in the coverage area. In order to improve the system receiving performance, feeder loss should be minimized.

(3) Determining the gain index of the repeater subsystem

The gain of the repeater subsystem is calculated according to the minimum SPFD and EIRP required by the system and the determined gain of the receiving antenna. Firstly, the upstream saturation or rated input power of the transponder is obtained through calculation, and then the gain of the transponder subsystem can be determined according to the output power allocated to the transponder subsystem during EIRP index decomposition. In general, the SPFD given in the payload system index has a range of variation, and therefore the total gain of the repeater subsystem also has a certain range of variation and can be adjusted by a program-controlled attenuator. For systems requiring compensation, the transponder should have automatic level control capability.

(4) Frequency response characteristic index decomposition

The amplitude–frequency and phase–frequency characteristics of the payload and communication quality mainly depend on the transponder. Although the non-ideality of the antenna's frequency response is small, it changes linearly within a single transponder channel, and the nonlinearity of the transponder in these aspects is always the most serious at the two edges of a transponder channel, so the effective load. The composite effect is equal to the linear superposition of the two. Therefore, when the frequency response characteristic index is allocated, the amplitude–frequency, phase–frequency characteristic and other indexes of the antenna within the single-channel bandwidth of any transponder should be given. The overall index of the effective load after deducting the index of the antenna is the transponder's corresponding indicators.

4.5 Communication Satellite Platform Design

The satellite platform is a combination of all service systems that support and guarantee the normal operation of the payload. According to the physical composition and service functions of the satellite system, the satellite platform can be divided into subsystems such as structure, thermal control, control, propulsion, power supply and distribution, measurement and control, and data management (or integrated electronics).

According to different space missions, satellite platforms that can be used directly or can be partially modified to support multiple payloads are called public satellite platforms. Public platforms generally have specific carrying capacity, power supply capacity, control capacity and computing processing capacity, generally keep the satellite configuration unchanged, and maintain the system, equipment configuration and interface of power supply and distribution, control, measurement and control, and data management (or integrated electronics) unchanged, keep the various interfaces with the carrier and launch site unchanged. The public platform puts more emphasis on multi-task and multi-load support capabilities, which can not only serve a certain range of load requirements but also have certain characteristics such as stability, independence and type spectrum. The use of satellite public platforms to develop satellites can inherit mature technologies, reduce development costs, shorten development cycles, improve system reliability and enhance satellite market competitiveness. It is suitable for the acceleration of satellite application development, especially the development needs of communication satellite industrialization. Therefore, we should try our best to choose public platforms for satellite development.

On the basis of clarifying user requirements and completing the preliminary demonstration of the satellite payload scheme, the selection of the satellite platform can be carried out. The factors that need to be considered for platform selection are: the weight of the payload; the power requirements of the payload; the installation area requirements of the payload's in-satellite equipment and the installation space and interface requirements of the extra-satellite antenna; the normal operation of the payload will control the satellite attitude and orbit special requirements for data management; requirements for measurement and control systems for satellite communications missions; determined types of launch vehicles, etc. The basic principles of satellite platform selection mainly include the carrying capacity and equipment layout space of the platform must be greater than the load demand; the power supply and adjustment capacity of the platform should not be lower than the load demand as much as possible; the technical status of the platform should be controlled to minimize the improvement of the design of the platform.

4.6 Communication Satellite System Design

Satellite configuration and layout design is an important overall design work. Its fundamental task is to determine the satellite's main load-bearing structure scheme, cabin composition, external dimensions and star-arrow interface under various constraints, and complete various satellite equipment layout design and analysis.

The configuration selection of the satellite needs to consider the way of satellite orbit (direct orbit itself), satellite orbit (sun-synchronous or geostationary orbit etc.), satellite on-orbit attitude (satellite sun exposure), satellite scale and carrying capacity factors such as the ability. According to the different main bearing methods, it can be generally divided into bearing cylindrical configuration, plate-frame configuration, truss configuration and hybrid configuration.

The load-bearing cylinder configuration is mainly suitable for large and medium-sized satellites in the middle orbit or high orbit itself (using a dual-element propulsion system). The load-bearing cylinder has good bearing conditions and can provide installation space for the storage tank, such as my country's DFH-3, DFH-4 platform, European Spacebus4000 platform, American LS-3000 platform etc. The plate-frame configuration is to form the main bearing structure with good mechanical conditions. Generally, more partitions need to be designed, which is more suitable for low-orbit remote-sensing satellites that can make full use of the partitions to install equipment, such as Lockheed Martin's A2100 platform. The advantage of the truss platform is that it has a simple structure and can provide more equipment layout space. It is more suitable for small and medium-sized satellites directly into an orbit, such as the French Poseidon platform and Galileo's networked navigation satellites. The service cabins with large-scale communication satellite platforms abroad also adopt a truss configuration, such as Boeing's BSS-601 platform and BSS-702 platform. The hybrid configuration is a combination of bearing tube, plate frame or truss structure. For example, the Eurostar 3000 platform is a bearing tube + plate-frame mixed main bearing configuration, and the Boeing 702 platform is a truss + plate frame hybrid main bearing configuration.

The satellite configuration design mainly considers the sub-cabin requirements, the installation requirements of special equipment (tanks, gas cylinders, related antennas etc.), structural strength and rigidity requirements and other factors. The satellite configuration and size need to be designed with comprehensive consideration of the following factors: the allowable diameter of the launch vehicle fairing, satellite payload capacity and platform service system capabilities, the number and size of antennas, the number and size of solar wing substrates and overall related layout requirements.

The shape of my country's DFH-4 satellite platform is a cuboid box structure. According to the guiding ideology of modularization and sub-cabin design, it is divided into two independent parts: the public platform structure and the communication cabin structure. The public platform structure can be further divided into two modules, the propulsion cabin and the service cabin according to its functions. Among them, the propulsion cabin structure is composed of the central bearing tube, the middle plate, the back floor and the propulsion cabin partition. It is a relatively independent and complete configuration state. It has a certain rigidity to maintain the configuration and accuracy. It is the skeleton of the entire satellite structure. The service cabin structure is composed of the north and south boards of the service cabin. After decomposition, the service cabin structure forms scattered structural boards, and each structural board needs a ground cage to provide support. The communication cabin structure consists of the floor, the north and south boards of the communication cabin and the north and south boards. The partitions are composed of a relatively independent "π"-shaped structure. Although the communication cabin structure itself has a certain degree of rigidity, it does not have the ability to maintain accuracy. The communication cabin holder needs to be used to provide support and accurate maintenance in the state of the communication cabin.

In order to achieve parallel development, parallel assembly and testing, the satellite configuration is generally divided into multiple cabins. For example, the DFH-4 communication satellite is divided into three parts: communication cabin, propulsion cabin and service cabin. Among them, the communication cabin is composed of a π-shaped structure consisting of the opposite floor, the communication cabin south/north plate and the communication cabin south/north partition; the propulsion cabin is composed of a central bearing tube, a back floor, a middle plate and the east, west, south, and the north consisting of four partitions; the service cabin is composed of the south and north instrument panels of the service cabin and the south and north battery panels of the service cabin.

The communication cabin is mainly installed with satellite payload equipment and tracking subsystem equipment; the propulsion cabin is mainly installed with oxidizer

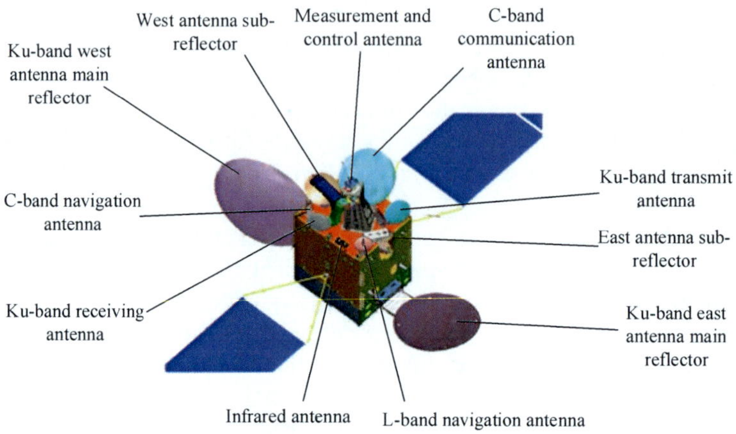

Fig. 4.5 Schematic diagram of communication satellite antenna composition

and combustion agent storage tanks (installed in the bearing cylinder), orbital engine, thruster, and main pipelines and control subsystems. The lower end of the bearing tube provides a mechanical interface with the launch vehicle; the service cabin is mainly installed with service system equipment such as control, power supply and distribution, measurement and control, and data management.

The layout of satellite antennas must first ensure the field of view requirements of each antenna, try to avoid stars blocking the antenna pattern, and mutual shielding between antennas. High-power antennas should avoid direct stars so as to avoid electromagnetic interference or scattering problems of the equipment in the satellite; antennas with the same frequency or working frequency band should be as far apart as possible or work opposite to each other; antennas that may interfere with each other should use different polarization methods; for moving antennas, the field of view needs to be analyzed at the extreme positions of each movement and ensure that the electromagnetic compatibility between the antennas meets the requirements. Take the DFH-4 platform satellite as an example. Considering the compatibility with the carrier fairing, the deployable antenna is generally installed on the east/west side of the satellite, and the fixed antenna or small-diameter movable antenna is installed on the opposite ground of the satellite. Figure 4.5 shows the main antenna layout of a communication satellite.

4.7 Flight Procedure Design

Satellite flight procedures refer to the satellite operating procedures designed in accordance with regulations during the process of launch vehicle take-off, separation of satellites and arrows, orbit change and normal satellite operation. The design should list the events and main functions in the operation of the spacecraft, the

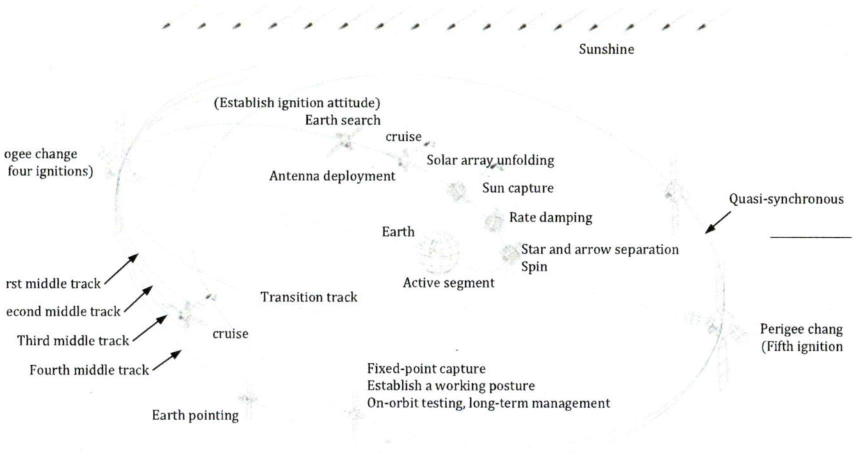

Fig. 4.6 Schematic diagram of flight procedures when a communication satellite is launched in a typical super-synchronous orbit

conditions required for the execution of each event, the execution sequence and start and end time of each event, the method of execution of each event and the result criterion.

The design and arrangement of flight procedures should consider the following requirements: (1) It must meet the restrictions of the launch vehicle and be compatible with the launch procedures of the launch vehicle; (2) The function limitations of the ground measurement and control network must be considered and compatible with the ground measurement and control network; (3) There are sequence requirements. The event must be arranged in strict accordance with the execution time of the event; (4) The important events arranged in the procedure should be left with a margin for the execution time, the execution method should be redundant and there should be repeated execution opportunities; (5) The procedure must be reduced. The degree of risk of execution improves the safety of program execution, increases the probability of completing tasks and improves the efficiency of program execution.

Figure 4.6 is a schematic diagram of a typical hypersynchronous orbit flight procedure of a GEO communication satellite. Table 4.2 gives examples of general flight procedures for GEO communication satellites.

Table 4.2 Typical flight events of communication satellites

No	Typical event	Function and effect
1	Star and arrow separation	Separation switch starts the on-board computer and the pyrotechnics manager program control, as the starting zero point for running the work program
2	Sun capture, establish cruise attitude	Capture the sun, eliminate the initial angular velocity and keep the satellite's attitude stable
3	Unlock and unfold the sun wing	The solar array starts to supply power to ensure sufficient power on the satellite
4	Unlock and unfold the antenna	Antenna is in the unfolded state
5	Active thermal control	Active temperature control of each component to ensure normal operation of the equipment
6	Capture the earth and establish a ground orientation attitude	Keep the earth in the field of view of the earth sensor and keep the satellite attitude stable
7	Establish an orbital ignition attitude	Adjust the yaw attitude and establish the ignition attitude
8	Satellite change	Carry out engine ignition according to orbit control strategy to change the orbit of the satellite from the initial orbit to the target orbit
9	Cut off the orbital engine fluid path and the propulsion subsystem gas path	After entering the quasi-synchronous orbit, the orbit-changing engine is isolated from the propulsion subsystem and the propulsion subsystem is changed from a constant pressure type to a drop pressure type
10	Fixed-point capture	Drift the satellite to a predetermined orbital position
11	On-orbit test	Establish the on-track test status according to the special test file, carry out various tests, evaluate the performance indicators of each subsystem and complete the delivery status setting
12	Track position retention	In accordance with the accuracy requirements of the satellite's north–south and east–west position maintenance, regular operations of the satellite's north–south position and east–west position

References

1. G. Maral, M. Bousquet, *Satellite Communications Systems*, vol. 2005, no. 4 (Wiley, 2009), pp. 36–40
2. ECSS. Multination design and test. ECSS-E-20–01A[R]. [S.I.]:ECSS, 2003. **2015**(11), 1–8
3. S. Cherry, Edom's law of bandwidth. IEEE Spectr. **7**, 58–60 (2004)
4. T. Kürner, S. Priebes, Towards THz communications-status in research, standardization and regulation. J. Infrared, Millimeter Terahertz Waves **1**, 53–62 (2014)
5. IEEE Draft Standard 802.11ad. Wireless LAN MAC and PHY Specifications-Amendment 4: Enhancement for Very High Throughput in the 60GHz Band. (2012)
6. D. Roddy, *Satellite Communications Third Edition* (McGraw-Hill, New York, 2001)
7. J. Feng, Z. Feng, Z. Wei, Load-aware offloading strategy in two-tier heterogeneous network. China Commun. **13**(8), 148–158 (2016)
8. J. Feng, S. Yang, Z. Feng, Vehicle-assisted offloading on metropolitan streets: enhancing geographical fluidity of wireless resources. IEEE Wirel. Commun. Lett. **6**(5), 622–625 (2017)
9. P.K. Wong, D. Yin, T.T. Lee, Analysis of non-persistent CSMA protocols with exponential back off scheduling. IEEE Trans. Commun. **59**(8), 2206–2214 (2011)

Chapter 5
Satellite System Spectrum Sensing

5.1 Introduction

With the rapid development of satellite communications, satellite channels have become more and more crowded, and the number of earth stations has increased dramatically, resulting in more serious signal interference and tighter spectrum resources. Satellite communication uses artificial satellites as relay stations to send radio signals and communicate between two or more earth stations. Satellite communication systems have the characteristics of long communication distance, wide coverage and high transmission quality. Satellite mobile channel transmission has the characteristics of openness, complex and diverse receiving environment, and random mobility of communication users, which determines that satellite mobile communication mainly has the following problems.

Firstly, the transmission characteristics of the satellite mobile channel determine that it is susceptible to multiple attenuation factors. Information sent by satellite to the ground station must pass through the atmosphere. During the propagation process, it will not only be subject to free space propagation, atmospheric refraction, ionosphere flicker, etc. but will be shielded by objects such as buildings and trees, causing the signal's propagation path to change dynamically. Due to the reflection and scattering of the signal, the phase and amplitude of the signal at the receiving end change, causing signal attenuation, which seriously affects the signal transmission efficiency and reliability.

Secondly, with the increase in the number of users and various multimedia services, the user's requirements for service quality continue to increase and the demand for satellite bandwidth resources continues to increase, resulting in increasingly prominent band resource issues. Service quality standards in satellite networks usually include bandwidth, throughput and real-time performance. Although some technologies can be used to help ensure service quality, sufficient frequency band resources are a prerequisite for service quality assurance for these high-speed transmission services, regarding the current ITU registration, the geostationary orbit C band communication satellite is nearly saturated, and the Ku-band communication

J. Zhang and J. Li, *Satellite Photoelectric Sensing Technology*,
https://doi.org/10.1007/978-3-030-89843-4_5

satellite is also very crowded. So how to improve bandwidth utilization has become the most urgent problem in the satellite communication field.

However, in contrast to the current situation that satellite communication frequency resources are becoming increasingly precious, the problem is that the spectrum resource utilization of satellite communication systems is low. Most fixed-satellite communication systems occupy frequency bands above 4 GHz. The actual test results show that the frequency spectrum utilization rate in this frequency band is very low. Therefore, the core of the shortage of spectrum resources in satellite communication systems is not the lack of sufficient spectrum resources, but the uneven use of spectrum resources. However, the reason for this phenomenon is the unreasonable spectrum allocation strategy. At present, there are many methods that can improve the utilization of unit spectrum resources, such as using multi-carrier frequency multiplexing or multi-antenna transmission technologies. The problem of uneven distribution of spectrum resources still exists. In order to fundamentally resolve this contradiction and improve the utilization of spectrum resources, people have considered the application of cognitive radio theory to satellite communication systems, and then proposed the concept of satellite cognitive wireless networks.

The background of satellite cognitive wireless technology is shown in Fig. 5.1. At present, the limited frequency resources allocated to fixed-satellite service (FSS) based on fixed bandwidth can no longer meet the growing demand for broadband multimedia services. Therefore, many scholars have begun to focus on the "dynamic" or "non-pre-coordinated" spectrum utilization. In satellite cognitive wireless networks, this dynamic spectrum access method means that cognitive users who have not been authorized by the frequency band access the authorized user's frequency band for communication without interfering with the authorized user, that is, the authorized user does not occupy the frequency band. In the case of cognitive communication, cognitive users use spectrum holes to communicate through dynamic spectrum access, thereby achieving spectrum resource sharing in satellite

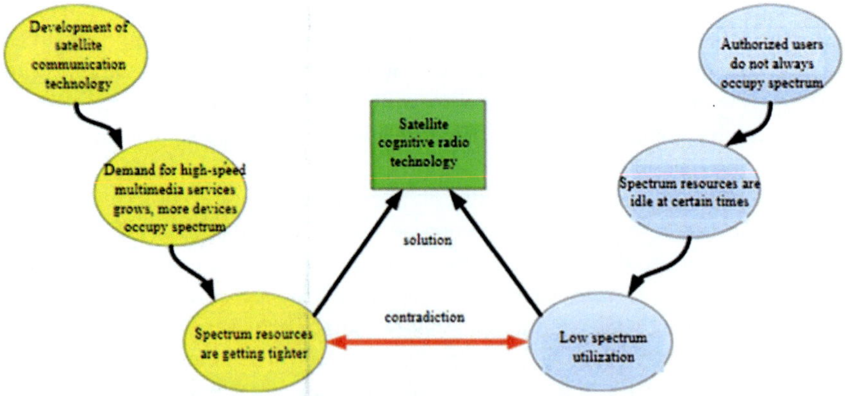

Fig. 5.1 Satellite cognitive radio technology proposed background

communication systems. The other aspect of dynamic spectrum access is that cognitive users will not coordinate with authorized users in advance, or inform authorized users of their own access behavior, and cognitive users cannot produce any form of communication behavior of authorized users. Interference in satellite cognitive wireless networks, for authorized users, cognitive users should be transparent [1].

The introduction of cognitive radio theory in satellite communication systems to realize the concept of dynamic spectrum access of satellite cognitive wireless networks is to solve the contradiction between the shortage of available spectrum resources and low spectrum utilization of satellite communication systems. According to the basic definition of cognitive radio, the problem of cognitive radio originates from the problem of signal detection, and spectrum perception is the cornerstone of cognitive space systems from theory to practical applications.

5.2 Spectrum-Sensing Concept

5.2.1 Spectrum Parameters

The effective use of frequency bands with sensing technology requires system designers and specifiers to define the parameters of the system's optional transmission methods. Too few parameter choices will limit the ability to perceive adaptation and too many parameter choices will increase the complexity of the system. A new concept of spectrum parameters was proposed in the report of the Spectrum Regulatory Commission. The Spectrum Regulatory Commission analyzed the benefits of using spectrum resources in terms of frequency, power, space and time. In the past, the Commission realized the three variables are expressed, and only the first three variables are taken into account when spectrum authorization is used. With the emergence and development of new technologies, the Commission considers adding time variables as the variable representation of spectrum resources, which is more conducive to dynamic spectrum allocation, and specifies special spectrum usage rights.

The new spectrum resource parameters are defined to increase the one-dimensional time based on the original frequency, space and energy so that the right to use the spectrum can be specified and allocated more dynamically. Existing intelligent perception systems are actually developed based on this definition [2].

The definition of the parameters of the spectrum resource is to show that the spectrum sensing technology can adjust all the parameters in the spectrum resource to make use of the frequency band more flexible. The latest digital signal processing technology and antenna technology can already realize the spectrum-sensing process in the entire spectrum space. Table 5.1 presents the definition of the spectrum parameters.

Table 5.1 Definition of spectrum parameters

Category	Parameter	Unit	Remark
Power	Power	W(or V/m)	Often used to independently represent spectral space
Frequency	Frequency	Hz	
Time	Time	Seconds	
Space	Location (longitude, latitude, altitude)	M	Three-dimensional space
	Signal direction (transmission direction, angle of arrival)	Rad	Two-dimensional space
Signal	Polarization (vertical, horizontal, left-handed circular polarization, right-handed circular polarization) encoding/modulation		Does not necessarily need to be orthogonal

For satellite communications, the power parameter of the spectrum can be used as the control variable of the transmitter's transmit power. The frequency parameter is used as the modulation frequency used by the transmitting antenna. The time parameter can provide a real-time reference for the second user's spectrum conversion. The spatial parameters determine which satellite the second user needs to target for communication. In summary, these spectrum parameters must be calculated by the sensing stations in the satellite cognitive network and stored in the candidate spectrum resource table as each element of a spectrum vector [4, 5].

5.2.2 *Spectrum-Sensing Technology Model*

1. **Definition of spectrum resources**

The goal of cognitive radio spectrum-sensing technology is to find the idle spectrum resources that are not fully utilized by the main user system at a specific time and place in the authorized frequency band that has been allocated to the main user system. This idle spectrum resource is called a spectrum hole and is also called a spectrum opportunity, as shown in Fig. 5.2. After the cognitive user discovers spectrum resources through spectrum sensing, they can access the spectrum opportunistically for communication. When the primary user appears and uses the idle spectrum, the cognitive user needs to exit the frequency band in time and look for new spectrum holes again.

2. **Multi-dimensional Spectrum Opportunities**

The definition of spectrum opportunity directly determines how to measure and detect the spectrum space. The traditional definition is that in a certain place, a frequency band that is not used at a particular time is called a spectrum opportunity.

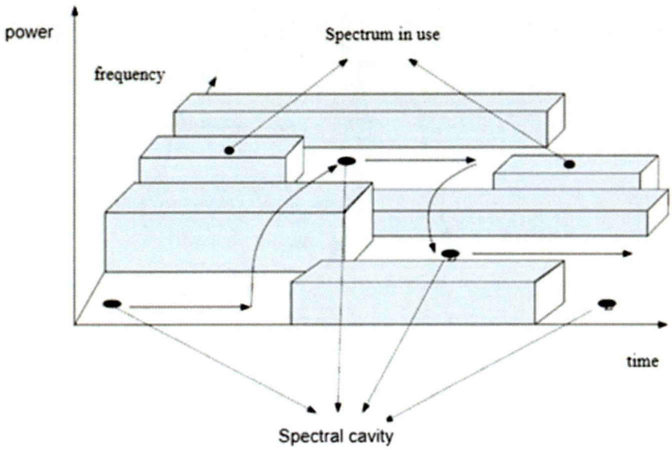

Fig. 5.2 Dynamic access of spectrum holes

This definition only considers three dimensions of the spectrum space: frequency domain, time domain and space domain. Traditional spectrum-sensing methods are only looking for spectrum opportunities in these three dimensions.

Beyond that, there are other dimensions where spectrum opportunities can be found. This section mainly introduces various situations where spectrum opportunities exist and the corresponding spectrum opportunity detection methods.

(1) Spectrum opportunities in the frequency and time domains

Spectrum opportunities in the frequency domain refer to frequency bands that have not been used in the frequency domain. The spectrum opportunity in the time domain refers to the unused frequency band in a certain period of time, as shown in Fig. 5.3. These two types of spectrum opportunities are common, and many works in the literature have discussed spectrum-sensing algorithms in the frequency and time domain.

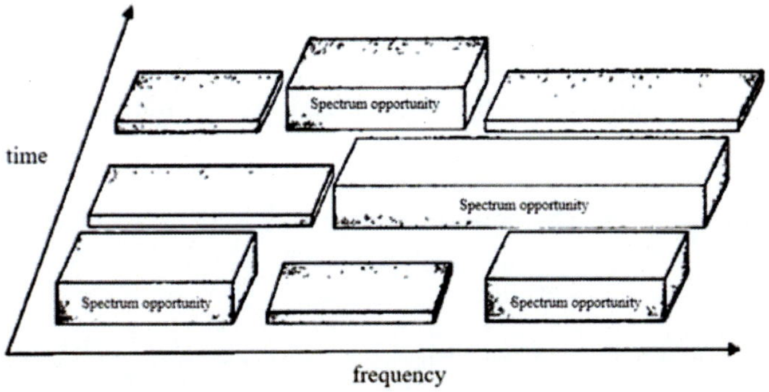

Fig. 5.3 Spectrum opportunities in the frequency and time domain

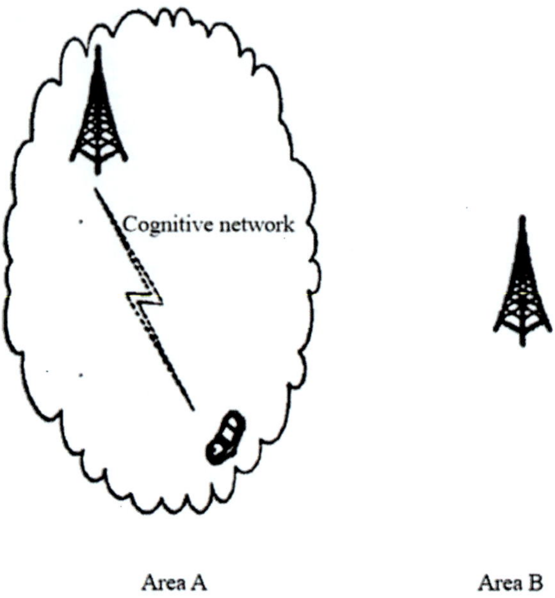

Fig. 5.4 Spectrum opportunities in airspace

(2) Spectrum opportunities in airspace

The spatial location includes the longitude, latitude and elevation of the base station and the distance to authorized users. In a certain area, there are some unused frequency bands. These free frequency bands can be regarded as airspace spectrum opportunities. As shown in Fig. 5.4, when the cognitive network is far from the authorized user, the cognitive user communicating in this area will not affect the remote authorized user due to path loss. However, it is worth noting that when exploring airspace spectrum opportunities, attention needs to be paid to the problem of hidden terminals.

(3) Spectrum opportunities in the coding domain

The spectrum opportunity in the coding domain refers to the use of an orthogonal code to the authorized user to communicate with the authorized user in the authorized user's area without interference to the authorized user, as shown in Fig. 5.5. Finding spectrum opportunities in the coding domain requires not only knowing the coding information of authorized users but also time and synchronization information.

(4) Angle of spectrum opportunities

When looking for angular spectrum opportunities, it is necessary to detect the beam direction of the authorized user, including azimuth and elevation, and the position of the authorized user. As shown in Fig. 5.6, after the beam direction and position of the authorized user are determined, the cognitive user can design the beam direction according to the location, avoiding the beam direction of the authorized user, and does not cause interference to the authorized user.

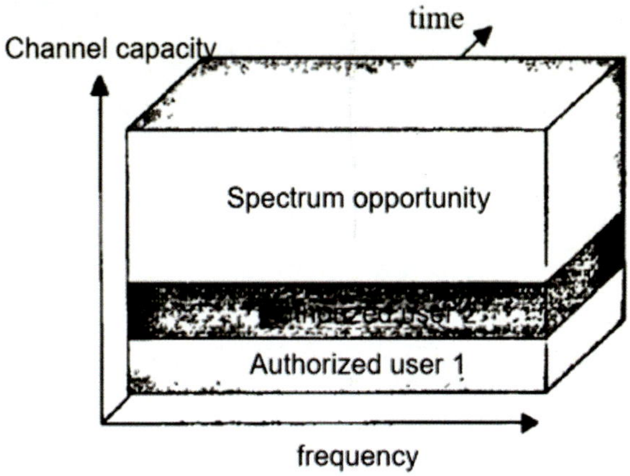

Fig. 5.5 Spectrum opportunities in the coding domain

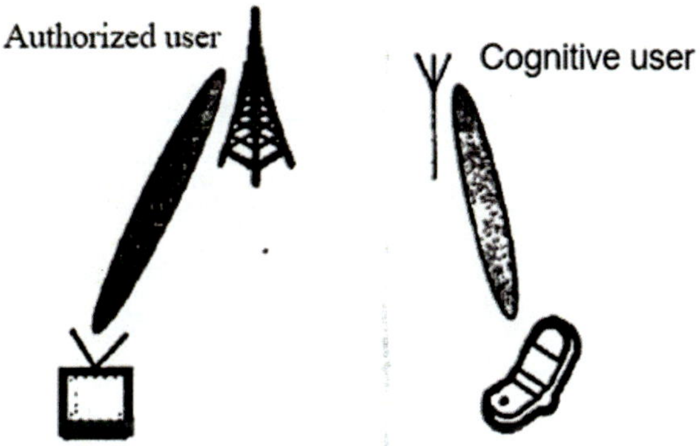

Fig. 5.6 Spectrum opportunity

It is very important to detect the spectrum opportunities in the different dimensions mentioned above so that it can provide cognitive users with more opportunities for spectrum use, and also improve spectrum utilization efficiency, and further alleviate spectrum resources caused by increased traffic shortage pressure.

5.2.3 Classification of Spectrum-Sensing Technology

At present, the spectrum-sensing technology is mainly divided into single-node sensing and multi-node cooperative spectrum sensing according to the user situation involved in sensing, as shown in Fig. 5.7.

Single-user spectrum sensing is divided into transmitter-based detection and receiver-based detection. Among them, the transmitter-based detection methods mainly include the matched filter method, the energy detection method and the cyclic stationary characteristic method, and the receiver-based detection methods mainly include the local oscillator leakage detection method and the interference

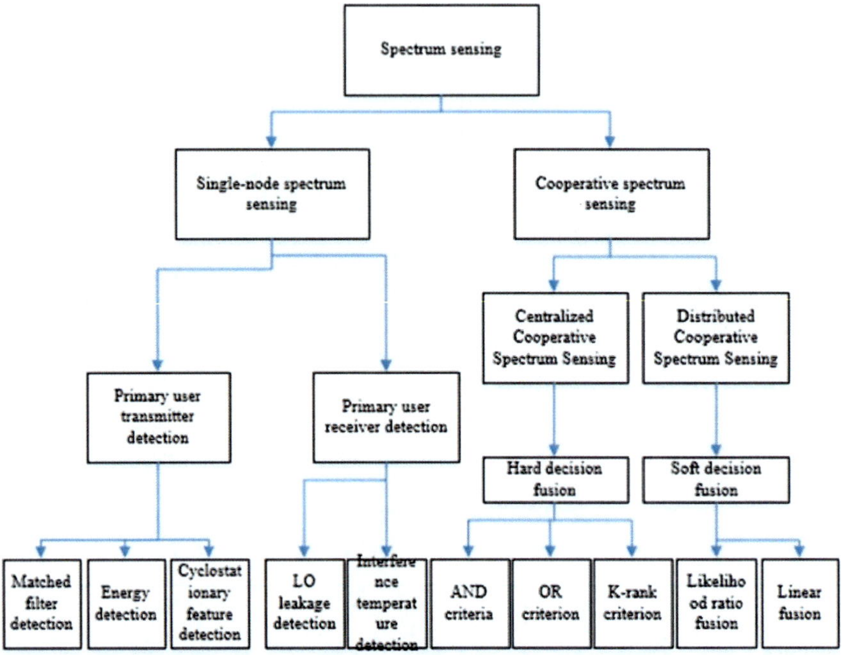

Fig. 5.7 Classification of spectrum sensing technology

temperature-based detection method. There are two types of cooperative spectrum sensing: centralized and distributed cooperative spectrum sensing. Centralized cooperative spectrum sensing is divided into soft decision fusion and hard decision fusion detection methods according to different data fusion methods. Hard decision fusion method is divided into AND criterion, OR criterion and K rank criterion method. The soft decision fusion method mainly includes likelihood ratio fusion and linear fusion method [3].

5.3 Spectrum-Sensing Principle of Space System

5.3.1 Typical Satellite Cognitive Network Scenario

At present, the application scenarios of satellite cognitive networks proposed by scholars at home and abroad are mainly divided into three categories: satellites use cognitive radio technology to share the free spectrum of the ground network; ground networks use cognitive radio technology to share the free spectrum of the satellite and satellite network utilization. Cognitive radio technology shares the free spectrum of other satellite networks [6–8].

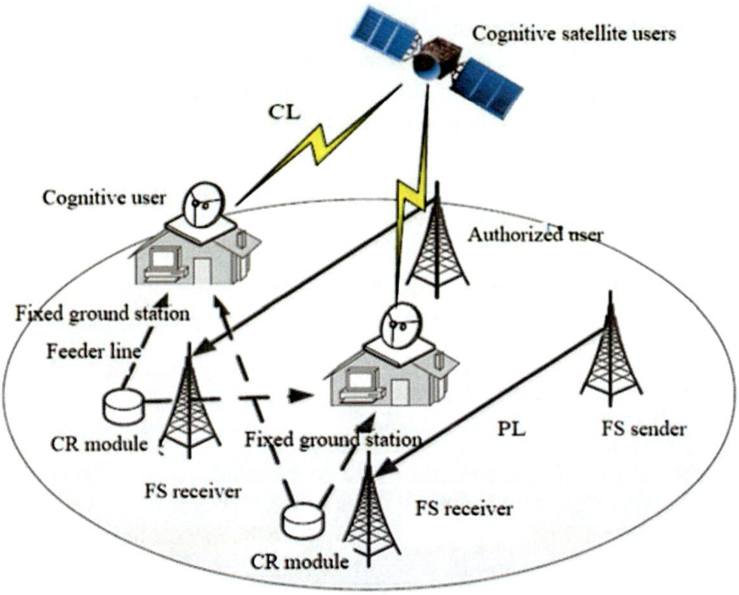

Fig. 5.8 Satellite networks share spectral holes in terrestrial networks using cognitive radio technology (scenario I)

1. **Satellite network uses cognitive radio technology to share ground network spectrum hole scenarios**

In this scenario where the satellite network uses cognitive radio technology to share the spectrum holes of the ground network (denoted as scenario I), the satellite communication system user acts as the cognitive system, and the users on the ground network are authorized users, and the satellite communication system will occupy the idle spectrum of the ground communication network communicates, as shown in Fig. 5.8.

The figure contains the primary users link (PL) and cognitive users link (CL). In the process of dynamic spectrum access of the satellite cognitive wireless network shown in scenario I, the FSS satellite terminals operating in the frequency band above 3 GHz and their corresponding fixed ground stations together form the cognitive network and the fixed service (FS). The network of users is an authorized network. The cognitive radio (CR) module is placed on the FS receiving end and senses whether the authorized user is using the licensed frequency band and feeds the perception result back to the cognitive system. When the cognitive network needs to establish an uplink communication link, the sensing result of the cognitive radio module will be fed back to the fixed ground station. If the fixed ground station finds available spectrum holes at this time, it performs dynamic spectrum access, and when the cognitive network needs to establish a downlink communication link, the sensing result needs to be fed back to the satellite through a high-gain antenna to inform the satellite node to use the spectral holes to transmit information.

In scenario I, the cognitive users in the cognitive network are only a small number of satellite nodes and their fixed satellite ground stations. The structure of the cognitive network is relatively simple and the spectrum resources that can be allocated are rich. However, due to the large satellite coverage area, the number of ground FS networks in the coverage area is large, that is, the number of authorized users is huge. This will make the spectrum-sensing process in cognitive radio systems very complicated. On the other hand, when the cognitive network needs to establish a downlink from a satellite to a fixed ground station, the CR module needs to feedback the sensing result to the satellite node through a high-gain antenna. Delay in cognitive radio network spectrum sensing is found.

2. **Terrestrial networks use cognitive radio technology to share satellite network spectrum hole scenarios**

Similarly, a ground wireless terminal equipped with a CR module can also make reasonable use of the spectrum holes in a satellite communication system to increase the utilization of spectrum resources by the network. The scenario where the terrestrial network uses cognitive radio technology to share the spectral holes of the satellite network (referred to as scenario II) is shown in Fig. 5.9.

In scenario II, the satellite communication system acts as an authorized user and the terrestrial wireless communication system acts as a cognitive user. Cognitive radio modules in the network are deployed near fixed-satellite ground stations. These modules will sense the use of authorized spectrum by authorized users, look for dynamic spectrum access opportunities and feed the perception results back to the

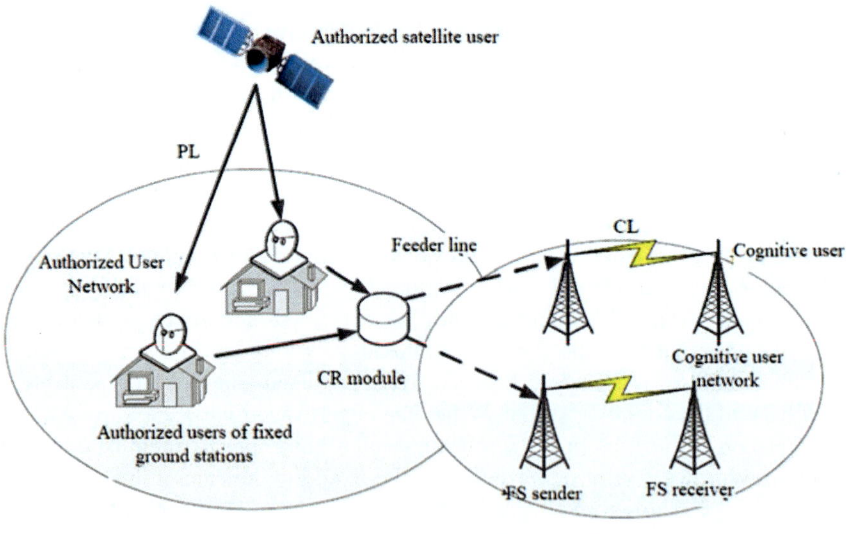

Fig. 5.9 Terrestrial networks share spectral holes in satellite networks using cognitive radio technology (scenario II)

cognitive system. Next, the users in the cognitive system will use these spectrum holes to communicate according to a certain dynamic spectrum access strategy.

The advantage of scenario II is that the cognitive network is set as a terrestrial wireless network, the construction cost is lower, the restrictions on the computing power and energy consumption of the cognitive user are smaller and the network is easier to maintain and optimize. However, the large coverage of authorized users in this scenario makes the number of cognitive users in the network huge and the structure of the cognitive network complex. On the other hand, the cognitive network in this scenario uses a distributed management architecture, which increases the complexity of cognitive network management and resource allocation.

3. **Satellite networks use cognitive radio technology to share other satellite network spectrum hole scenarios**

In another application scenario of dynamic spectrum access of satellite cognitive wireless network, satellite nodes equipped with CR modules share spectrum holes of other satellite networks, and satellite networks use cognitive radio technology to share spectrum hole scenarios of other satellite networks (recorded as scenario III), as shown in Fig. 5.10.

In this scenario, cognitive users and authorized users are nodes in the satellite communication system. The cognitive satellite users in the cognitive network can use the sensing results deployed in the authorized user network to make decisions and dynamically access the authorized satellite users, and to communicate with each other so as to effectively use spectrum resources without interfering with communication between authorized satellite users and their corresponding ground stations.

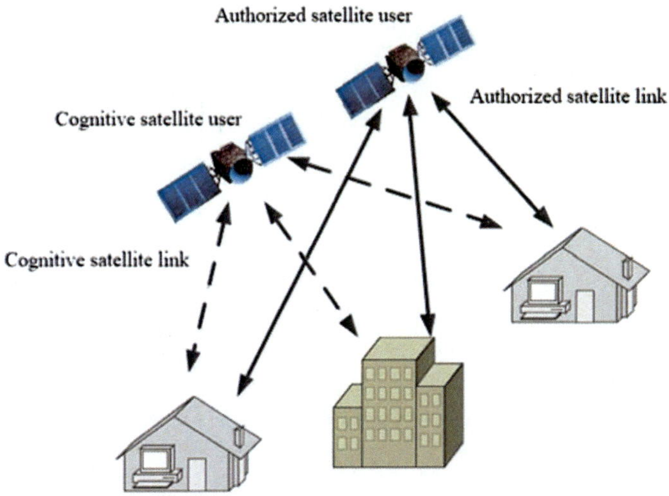

Fig. 5.10 Satellite networks share cognitive holes in other satellite networks using cognitive radio technology (scenario III)

Although scenario III has fewer cognitive users and more resources can be allocated, this scenario sets the satellite node as a cognitive user, which places high requirements on the computing power and energy consumption of the satellite. In addition, in this scenario, the CR module deployed in the authorized user network needs to sense the spectral holes in its satellite-to-ground link and feedback the sensing results to the satellites and their ground stations in the cognitive user network. This undoubtedly greatly increases the difficulty of the spectrum-sensing process and causes a delay in spectrum sensing.

Through comparison and analysis, the advantages and disadvantages of the above three types of satellite cognitive wireless network application scenarios can be summarized and is shown in Table 5.2.

Through the analysis of the characteristics of the dynamic spectrum scene of a typical satellite cognitive wireless network, it can be found that the use of cognitive radio theory to discover and share the spectrum holes in satellite communication networks can make the spectrum more efficient and can effectively alleviate the

Table 5.2 Dynamic spectrum access scenarios and characteristics of typical satellite cognitive wireless networks

Dynamic spectrum access scenario for satellite cognitive wireless network	Advantages	Disadvantages
Satellite networks use cognitive radio technology to share ground network spectrum holes (scenario I)	1. Low number of cognitive users 2. Wide coverage 3. Abundant spectrum resources are available	1. Large number of authorized users makes spectrum sensing difficult 2. There is a delay in the spectrum sensing of the cognitive network, and the construction cost of the CR module in the cognitive network is high
Terrestrial networks share satellite network spectrum holes using cognitive radio technology (scenario II)	1. Cognitive users are on the ground, facilitating network maintenance and optimization 2. Relatively small restrictions on cognitive user processing power and energy consumption	1. Coverage of authorized users is large, the number of network users is large and the structure is complex 2. Using a distributed management architecture, it is difficult to optimize the allocation of network resources and suppress interference
Satellite networks use cognitive radio technology to share spectrum holes in other satellite networks (scenario III)	1. Simple cognitive wireless network construction 2. Fewer users 3. Wide coverage	1. Satellites as cognitive users have limited processing power and energy consumption, and high construction and maintenance costs 2. The spectrum-sensing process is difficult and time-consuming

current existence. There is a conflict between the low utilization of spectrum resources and the shortage of spectrum resources. However, in this process, satellite nodes' energy consumption and computing power limitations have become important factors restricting the application of cognitive radio technology to satellite communication networks.

5.3.2 Problems in Satellite Spectrum Sensing

The problem of applying cognitive radio technology to satellite communication systems has become a research hotspot in the field of satellite communications at home and abroad in recent years. In response to this problem, domestic and foreign scholars have done some theoretical research and simulation verification of related technologies. However, there is currently no perfect application of cognitive radio technology in satellite communication systems. With the development of related research, the application of cognitive radio will not only have basic spectrum awareness but also have more intelligent functions and attributes [9, 10]. The problem of low practical spectrum utilization in satellite communication systems has become a prerequisite for the introduction of cognitive radio technology. However, on the other hand, satellite communication systems have higher requirements for the stability and flexibility of communication links, so they have a sense of spectrum in space systems. The application of technology also requires the following aspects:

(1) The channel conditions in satellite communication systems are special. There is a Rice channel in a satellite communication system and the signal-to-noise at the receiving end is relatively low. Traditional cognitive radio spectrum-sensing algorithms are suitable for a single sensing node, such as the matched filter detection method, energy detection algorithm, cyclic stationary detection algorithm, etc. Under the condition of satellite communication channel, it is easy to be interfered, and the algorithm performance is degraded. Therefore, to use a centralized multi-node spectrum sensing method in satellite cognitive wireless networks to improve the system's spectrum sensing performance so as to realize the fast and accurate perception of the communication behavior of authorized users, it is necessary for satellite cognitive wireless networks to face the key issue. Therefore, it is necessary to study the spectrum-sensing algorithms suitable for satellite cognitive wireless network scenarios. At present, there are few studies in this area at home and abroad.

(2) The compromise between perceived effectiveness and reliability. In order to improve the utilization of spectrum resources of cognitive systems, it is hoped that various resource consumptions such as time, energy consumption and sampling overhead for spectrum sensing can be minimized so that more resources can be used for effective data transmission. At the same time, in order to accurately detect spectrum opportunities, it has to spend more time, energy consumption, sampling overhead and other resources on spectrum detection to improve detection performance. Therefore, spectrum-sensing technology needs to study how to achieve a reasonable compromise between effectiveness and reliability.

(3) Detection of wideband weak signals. In satellite star group network communication, different satellites and different payloads of the same satellite work in different frequency bands due to different missions, and the frequency spectrum is wide. In order to effectively use the spectrum holes in the widest possible frequency band, cognitive users need to have a wide spectrum detection range and high detection sensitivity. However, due to the hardware conditions of the device, complex wireless propagation environments (such as wireless fading and the impact of noise) and other constraints, how to achieve reliable and effective detection of weak signals in a wide frequency band in a hardware-constrained and complex wireless propagation environment is a problem faced by spectrum-sensing technology.

References

1. T. Yucek, H. Arslan, A survey of spectrum sensing algorithms for cognitive radio applications. IEEE Commun. Surv. Tutorials **11**(1), 116–130 (2009)
2. F. Capar, I. Martoyo, T. Weiss, F. Jondral, Comparison of bandwidth utilization for controlled and uncontrolled channel assignment in a spectrum pooling system, in *IEEE 55th Vehicular Technology Conference*, vol. 3 (Spr. 2002), pp. 1069–1073
3. R. Tandra, A. Sahai, Fundamental limits on detection in low SNR under noise uncertainty, in *2005 International Conference on Wireless Networks, Communications and Mobile Computing* (2005), pp. 464–469
4. A. Ghasemi, E.S. Sousa, Optimization of spectrum sensing for opportunistic spectrum access in cognitive radio networks, in *Proceedings of 4th IEEE Consumer Commun. Networking Conference (CCNC)* (2007), pp. 1022–1026
5. Y.-C. Liang, Y. Zeng, E. Peh, A.-T. Hoang, Sensing-throughput tradeoff for cognitive radio networks, in *Proceedings of IEEE ICC* (2007), pp. 5330–5335
6. Y.-C. Liang, Y. Zeng, E. Peh, A.-T. Hoang, Sensing -throughput tradeoff for cognitive radio networks. IEEE Trans. Wirel. Commun. **7**(4), 1326–1337 (2008)
7. L. Tang, Y. Chen, E.L. Hines, M.-S. Alouini, Effect of primary user traffic on sensing - throughput tradeoff for cognitive radios. IEEE Trans. Wirel. Commun. **10**(4), 1063–1068 (2011)
8. Y.-Y. Pei, A.-T. Hoang, Y.-C. Liang, Sensing -throughput tradeoff in cognitive radio networks: how frequently should spectrum sensing be carried out?, in *IEEE 18th International Symposium on Personal, Indoor and Mobile Radio Communication* (2007), pp. 1–5
9. D. Xue, X. Wang, E. Hossain, Optimization of periodic channel sensing by secondary users in a cognitive radio network, in *IEEE Global Telecommunications Conference* (2010), pp. 1–5
10. S. Geirhofer, L. Tong, B.M. Sadler, Cognitive radios for dynamic spectrum access– dynamic spectrum access in the time domain: modeling and exploiting white space [J]. IEEE Commun. Mag. **45**(5), 66–72 (2007)

Chapter 6
RF Channelization Technology

6.1 Research Background of Broadband RF Channelization Receiving Technology

With the development of communication technology, the world has entered the era of information explosion. Whether it is common civilian broadband multi-service access, communication navigation identification and other technologies, or electronic countermeasures, spectrum monitoring and communication intelligence technologies in information warfare, the rapid development of short-range transmission and processing of narrowband signals can no longer meet modern communication requirements. Application requirements are long-distance transmission of broadband signals, high-performance sensing, reception and processing. In recent years, with the development of frequency-hopping confidential communications and high-frequency agile radar, the electromagnetic bandwidth required for transmission processing in military communications has increased significantly. It is required to monitor the bandwidth from GHz to dozens of GHz to achieve real-time detection of information in a wide spectrum, collect, alert and publish. At the same time, as the working frequency of the radar signal extends down to the short-wave band and the working frequency of communication extends to the millimeter-wave band, the signals of the two overlap in the spectral distribution, and there is also aliasing in the signal system. It is distinct from the traditional concept. It is difficult for a single electronic countermeasure system to comprehensively and effectively suppress the enemy with high technology, and it is impossible to obtain a strong information acquisition advantage and electronic offensive power. Driven by demand traction and technology drive, high-performance electronic equipment is moving toward integration [1]. While reducing equipment resource redundancy, it can also dynamically configure hardware resources to maximize system information detection and processing capabilities [1]. Therefore, ultra-wideband, multi-band integrated radar, communication and electronic warfare multifunctional integrated radio frequency system is an inevitable trend for future military and civilian applications.

J. Zhang and J. Li, *Satellite Photoelectric Sensing Technology*,
https://doi.org/10.1007/978-3-030-89843-4_6

A high-performance RF system must have large instantaneous detection bandwidth, high resolution, large dynamic range, and the ability to detect and receive distortion-free and real-time processing of multiple frequency and multi-form signals that arrive at the same time. It can be seen that the real-time capture of broadband agile information is the core enabling technology for many application requirements. With the development of RF microwave technology, various RF receiving technologies have been proposed and widely used [2–4]. The superheterodyne receiving technology converts the high-frequency signal received by the antenna into a fixed intermediate frequency signal after amplification and down-conversion and then performs the next step of frequency conversion or demodulation. It is the most widely used narrowband RF receiving system structure. However, this technology is severely disturbed by the image frequency. Multiple high-performance filters are required to suppress the image frequency, which greatly reduces the stability and integration of the system. Image suppression receiving technology uses the orthogonal suppression method to achieve image suppression through the mismatch of amplitude and phase. However, the accuracy deviation of the amplitude and phase difference of the quadrature signal seriously affects the performance of the system. Zero-IF reception technology directly down-converts useful signals to baseband, which avoids image interference and avoids the use of additional microwave devices, which simplifies the circuit module. However, the DC offset introduced by each module and the DC offset introduced by the unsatisfactory isolation of each port will be directly superimposed on the signal, causing interference.

As the communication frequency increases and the bandwidth increases, facing a high-density and complex signal environment, the structure of a traditional receiver, as shown in Fig. 6.1, can only intercept signals at a single frequency point. For the simultaneous detection and processing of point signals, various broadband radio frequency receiving mechanisms were proposed in the 1970s and 1980s, such as crystal video reception, compression reception, instantaneous frequency measurement reception and channelized reception [5–8]. The channelized receiver divides the signal to be received into channels in the frequency domain through a filter, divides a wideband signal into multiple narrowbands, and perceives and receives signals in different bands in parallel and real time. This receiving mechanism has a nearly 100% interception rate for signals and has the ability to process multiple signals simultaneously, meeting the needs of large instantaneous detection bandwidth, with

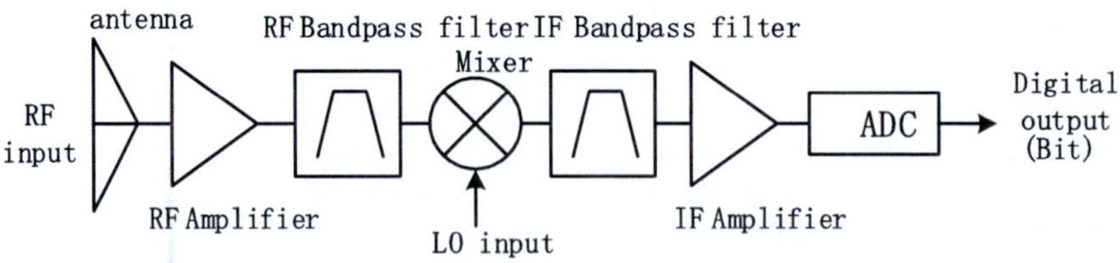

Fig. 6.1 Basic structure of a traditional RF receiving system

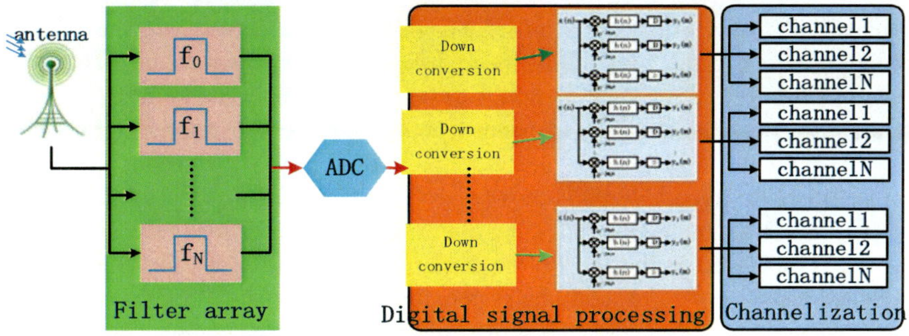

Fig. 6.2 The basic architecture of a digital channelized receiving system

high sensitivity and frequency resolution, and strong anti-interference ability. A wideband RF receiving structure uses and meets a variety of military demand bandwidths. The analog channelized receiver is composed of microwave devices. The channel is poorly balanced, and the power consumption, volume and cost of the system increase with the number of channels. With the development of digital signal processing technology and the increase of the sampling rate of analog-to-digital converters, since the 1990s, broadband digital channelization receiving technology has combined the advantages of channelization and digitization, and gradually replaced analog channelization as a new [9]. The basic structure is shown in Fig. 6.2. The core of digital channelization is the structural design of the digital filter array. Digital circuits implemented using various integrated chips can flexibly divide and switch channels, with high-frequency resolution and good system stability. Digital channelization reception machines play an important role in electronic countermeasures.

With the increasingly harsh military environment such as electronic countermeasures, many electronic bottlenecks of traditional microwave circuit systems can no longer be ignored, which seriously affects the improvement of system performance. First of all, current RF devices have great losses in high-frequency bands, and cannot support the processing of ultra-wideband, multi-band RF signals, challenging the ability to extract weak power signals in a noisy background. Although the use of different devices for different frequency bands can increase the bandwidth, the isolation and leakage between the internal components of the system module cannot be ignored, and the participation of many devices will also lead to an increase in system size, weight and power consumption. In addition, high-power clutter and interference signals in the signal environment under the nonlinear effects of saturation, mixing, frequency doubling and intermodulation of the RF device will cause serious distortion of the signal, leading to errors and misjudgments in receiving information, and don't cope with strong interference electromagnetic environment. Secondly, coaxial cables and waveguides are also one of the significant limiting factors for channelized RF transmission and processing. Traditional electronic warfare receivers use coaxial cables as the transmission medium for radio frequency, and their anti-electromagnetic interference capability is relatively weak. They are severely affected by climate change. They cause huge losses for long-distance transmission of high-frequency signals, with losses up to 360 dB/km @5.8 GHz. It will also increase with the increase of radio frequency, thereby limiting the high-frequency expansion of the system. At the same time, coaxial cables are bulky, heavy and have poor flexibility and reconfigurability. In addition, ultra-wideband channelized reception requires high-fidelity digital bandwidth. Digital signal processing is used to mine target information from it. The increase in bandwidth is also a great challenge for high-speed processing chips, high-rate sampling modules and high storage memory. Cost mass production cannot be achieved in the short term. It can be seen that the performance of the wideband RF channelized receiving system that processes signals on the traditional electrical domain is limited, and it is difficult to cope with the complex electromagnetic environment and military and civil needs. With the development of optical fiber technology in recent years, microwave photonics, which combines microwave technology and photonics technology, has shown great advantages and potential in the transmission, sensing and processing of broadband RF signals.

6.2 Channelized Receiver

Channelized receivers first appeared in the 1970s and 1980s and refer to the technology of dividing the frequency domain into multiple channels and then measuring them separately. The first to appear is an analog channelized receiver, which is considered to be the most suitable reconnaissance receiver for use in electronic warfare due to its higher reliability and intercept probability. However, the analog channelized receiver system has many disadvantages: the system structure is complex, the power loss is too high, the receiver volume is generally large and the production cost is high. With the development of digital signal processing technology, the development of digital channelized receivers becomes possible. The digital channelization receiver digitally stores the analog signal after AD conversion and processes the signal through software or programmable digital devices, which greatly reduces the complexity and power consumption of the system. Analog channelized receivers are mostly uniform channels. The commonly used forms are pure channelized receiver, folded channelized receiver and time-division channelized receiver.

Pure channelization is implemented in a hierarchical manner, as shown in Fig. 6.3; that is, a multi-level frequency splitter is used to convert the broadband signal multiple times to change the RF signal into multiple signals with the same intermediate frequency and bandwidth. Then perform signal detection on each signal to obtain the frequency domain parameters of the signal. The composition block diagram is shown in Fig. 6.3.

Pure channelization uses a large number of frequency splitters and mixers, which greatly increases the complexity of the system. Folding channelization has improved it, combining multiple splitters in the second stage into one, and folding the final signals of multiple channels into one channel, and combining the output of the two-stage splitter to get the signal frequency information. It can be known from the above principle that the folded channelization receiver also superimposes noise when

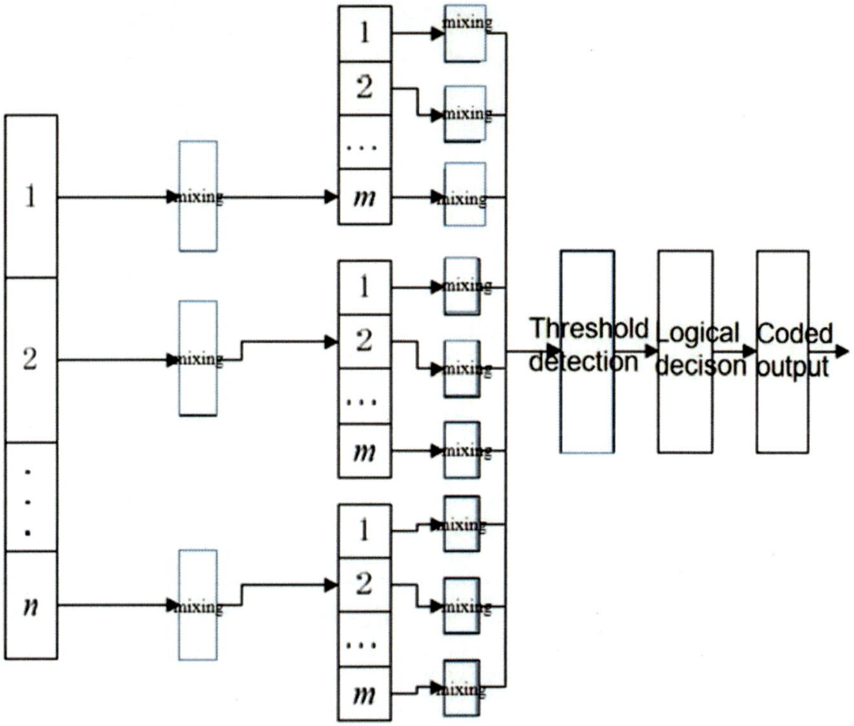

Fig. 6.3 Structure of a pure channelization system

performing signal folding, which reduces the detection performance of the receiver. When there are signals of different frequencies arriving at the same time, the folded channelization process generates a frequency measurement blur. The composition block diagram is shown in Fig. 6.4.

When the folded channelized receiver is simplified in the structure of the channelized receiver, the noise is accumulated, which reduces the detection performance of the system. In the time-division channelization receiver, a channel with a signal is connected to the second-stage splitter for processing before the second-stage splitter, and other channels are disconnected. This method avoids the superposition of noise, but it also cannot perform frequency domain analysis on signals in different channels that arrive at the same time.

Non-uniform channelization appeared after the advent of digital signal processing technology and is mainly used in the reconnaissance of communication signals. Since the distribution of electromagnetic signals in the frequency spectrum is non-uniform under actual conditions, the non-uniform channelized receiver solves the problem of resource waste caused by the uniform channelized receiver. With some prior knowledge, non-uniform channelized receivers use narrower bandwidth filters to improve measurement accuracy in frequency bands with dense communication signals and use wider bandwidth filters to expand monitoring scope with sparse frequency bands

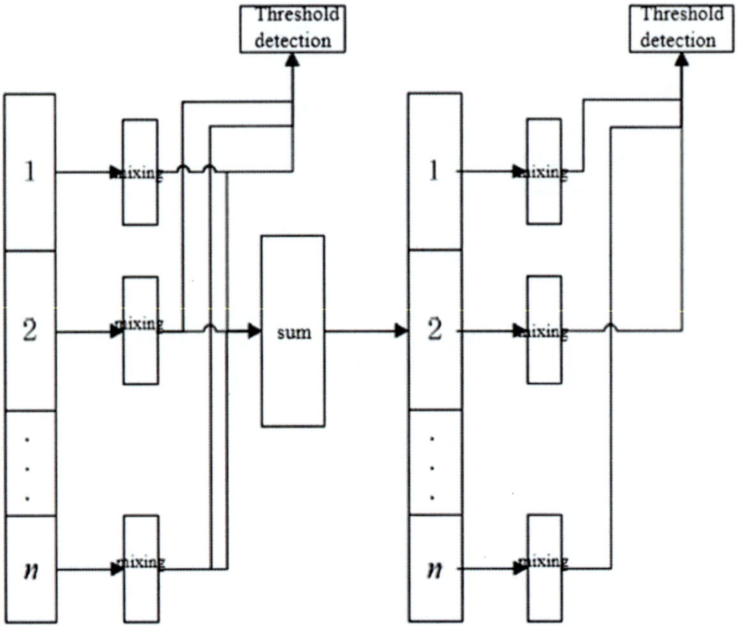

Fig. 6.4 Structure of a band-folding channelization system

communication signals. Traditional channelization implementations often use fixed channels and cannot dynamically adjust parameters based on the electromagnetic environment.

6.2.1 Analog Channelized Receiver

From the perspective of technical principles, analog receivers can be mainly divided into crystal video receivers, superheterodyne receivers, instantaneous frequency measurement receivers, compression receivers and channelized receivers.

1. **Crystal video receiver**

Crystal video receiver is the simplest electronic warfare reconnaissance receiver, as shown in Fig. 6.5. It is tuned by band-pass filtering the RF signal [10]. The filtered RF signal is input to a crystal detection diode to be converted into a video pulse sequence. Then it is detected by a logarithmic video amplifier and threshold. When the threshold voltage is specified, a signal is considered to exist in this frequency band. The crystal video receiver uses a tunable band-pass filter to achieve frequency selection, so the frequency resolution of the entire crystal video receiver is determined by the bandwidth of the band-pass filter. In addition, the filtering characteristics of analog band-pass filters are greatly limited at high frequencies, and their

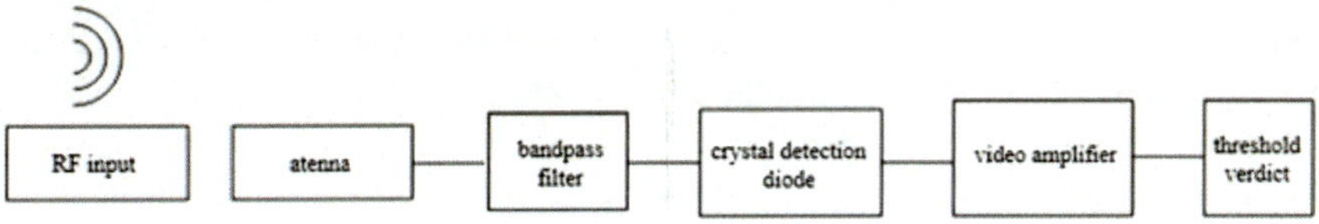

Fig. 6.5 Schematic diagram of crystal video receiver

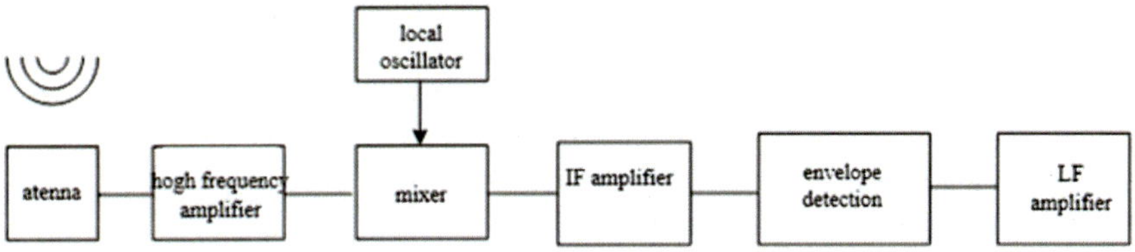

Fig. 6.6 Basic structure of a superheterodyne receiver

high-frequency losses are large. Therefore, crystal video receivers cannot be used in high-frequency electromagnetic detection environments. At the same time, the instantaneous bandwidth of this type of receiver is very narrow, and it cannot realize the real-time detection of multiple signals in a wide range. Therefore, this type of receiver cannot achieve a 100% intercept probability. The crystal video receiver can only distinguish the amplitude information of the target signal, but cannot detect the phase information of the target signal. Therefore, this type of receiver can only receive the target signal with a simple amplitude, and cannot distinguish multiple target signals that arrive simultaneously within the filter bandwidth. It is only used in some simple scenarios, such as low signal density and simple signal format scenes.

2. **Superheterodyne receiver**

The basic principle of a superheterodyne receiver is shown in Fig. 6.6. The signal received from the antenna is amplified by a high-frequency amplifier, and then the amplified signal and the local oscillator signal are input into a microwave mixer. Through mixing, the signal is down-converted to the intermediate frequency, and then the intermediate frequency signal is amplified by the intermediate frequency amplifier. Finally, the intermediate frequency signal is subjected to envelope detection to obtain a low-frequency signal, thereby realizing detection of the high-frequency signal. By changing the local oscillator frequency, the superheterodyne receiver can detect and receive radio frequency signals in a wide frequency range. At the same time, this type of receiver completes the detection of the signal in a narrower IF domain. Except that the tuning circuit of the high-frequency amplifier and the antenna loop needs to be unified with the resonant circuit of the local oscillator, the amplifiers at the intermediate frequency and the filters are relatively fixed. So when receiving different frequencies, there is no need to connect the intermediate frequency amplifier, and the detection receiver module is changed later. Thus the tuning function can be easily implemented. Therefore, this type of receiver can achieve better frequency and gain tuning. In addition, this type of receiver can achieve higher sensitivity and better frequency selectivity. Now superheterodyne receivers can be divided into the following types according to their functions: narrowband superheterodyne receivers, wideband superheterodyne receivers and wideband preselected receivers. In a superheterodyne receiver, the IF signal that passes through the mixer passes through an IF filter. This filter directly determines the detection sensitivity, frequency resolution and interception probability of the entire receiver. Narrowband superheterodyne receiver refers to the narrow bandwidth of the IF filter, which can achieve high-resolution signal detection, and the sensitivity of the narrowband superheterodyne receiver is relatively high. However, high detection sensitivity and resolution come at the expense of the local oscillator scanning time. The direct result is that the scanning time of this type of receiver is longer, the interception rate is not high and the detection bandwidth is greatly limited. At present, the bandwidth of a narrowband superheterodyne receiver can generally cover the instantaneous bandwidth of a single radar signal, and its typical instantaneous bandwidth value is 20 MHz. In order to overcome the shortcomings mentioned above, broadband superheterodyne receivers have emerged at the historic moment, and their instantaneous bandwidth can generally reach 200–1000 MHz, which can realize the detection, reception and processing of agile frequency conversion signals and broadband modulation signals. At the same time, this type of receiver can shorten the local oscillator scanning time and increase the interception probability of the entire receiver. But this type of receiver has to face the disadvantages of lower sensitivity and lower frequency resolution. The wideband preselection receiver suppresses various parasitic interference between channels by using a wideband preselection filter and unified tuning of the local oscillator, thereby improving the linearity of the entire receiver and achieving high-fidelity signal reception and processing [11].

3. **Instantaneous frequency measurement receiver**

The instantaneous frequency measurement receiver uses the method of delayed auto-correlation to convert the frequency information in the signal to be measured into power information and establishes the relationship between frequency and power, thereby achieving the measurement of the frequency of the signal to be measured. Instantaneous frequency measurement receiver is mainly composed of the following parts: limiter amplifier, phase detector and encoder. As shown in Fig. 6.7, we can see that the entire system is composed of a combination of a delay line with a delay of

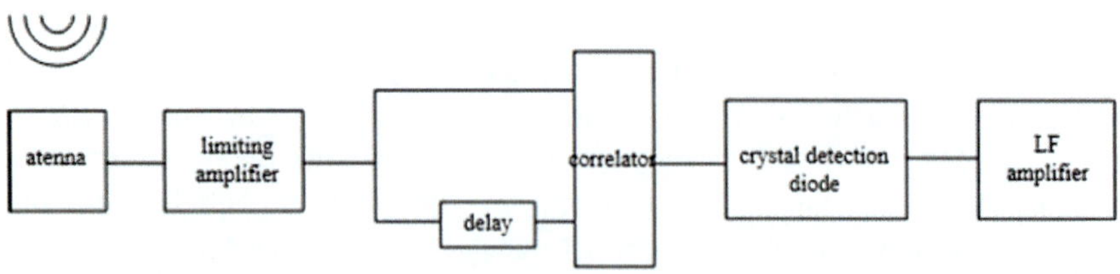

Fig. 6.7 Basic structure of instantaneous frequency measurement receiver

τ and a correlator. After passing the correlator, an automatic test signal with a delay of τ is generated correlation function. By analyzing the autocorrelation function, the relationship between frequency and power can be determined so as to determine the frequency of the input signal according to the change in power. The main advantages of the instantaneous frequency measurement receiver are the small size of the entire system, fast frequency measurement, high accuracy and wide instantaneous bandwidth. The measurement accuracy of a typical wideband instantaneous frequency measurement receiver is now 3–5 MHz. Instantaneous frequency measurement receivers have been widely used in electronic warfare detection and collection systems such as signal detection and jammer frequency guidance. However, the bandwidth of this type of receiver is still limited by analog devices, and it is difficult to achieve a large instantaneous bandwidth. At the same time, it is difficult for this type of receiver to measure multiple signals in real time. Simultaneously, frequency blurring and other phenomena will occur during the frequency measurement, which makes it difficult to apply in a high signal density environment. Many scientific researchers are now optimizing the instantaneous frequency measurement receiver so that it can process multiple simultaneous signals at the same time. But the limiting amplifier used in instantaneous frequency measurement is a fundamental limitation. As we all know, the limiter is a nonlinear device. When only one signal is incident, the nonlinear effect of the limiter on the signal manifests itself: the harmonic component of the single-frequency signal will appear at the output of the limiter. Moreover, these components can be suppressed by adding a band-pass filter. However, for a two-tone signal incident on the limiter, the third-order intermodulation distortion produced by it is difficult to be suppressed by the filter so that the output of the correlator is no longer the autocorrelation function of the multi-signal that we want. There will be many nonlinear distortions in the middle, resulting in inaccurate final measurement results [12].

4. **Compression receiver**

The compression receiver is to transform the input signal into short pulses in the time domain and establish the relationship between the frequency of the signal to be measured and the time of the pulse so as to measure the frequency of the signal to be measured. The compression receiver is mainly divided into the following parts: low-noise amplifier, chirp signal source, mixer, dispersive medium and real-time spectrum analysis unit. As shown in Fig. 6.8, the signal received by the antenna first passes through a low-noise amplifier, and then the low-noise amplified signal and the linear chirp signal are multiplied in the mixer. The modulated signal is then compressed into a short pulse through a dispersive medium. Finally, Fourier analysis is performed on these short pulses to achieve the conversion of frequency information to time information. The advantage of a compression receiver is that it can detect multiple RF signals with high sensitivity and a large dynamic range in real time. The receiver can acquire the time domain information of the signal and the frequency domain information of the signal. Therefore, the receiver can be applied to a signal-dense complex environment with agile frequency. However, this type of receiver has

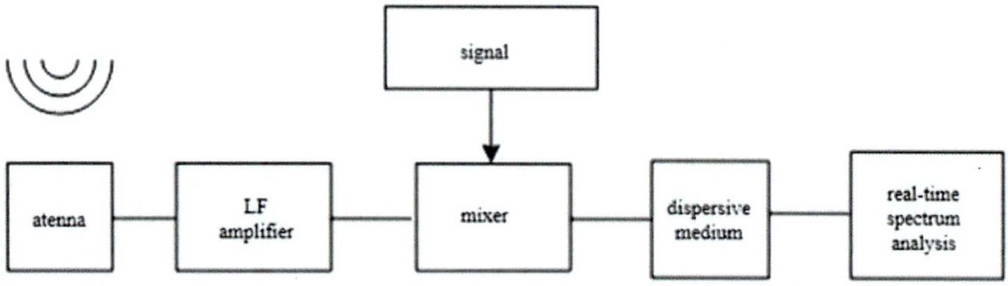

Fig. 6.8 Basic structure of a compression receiver

a narrow instantaneous bandwidth, which makes it difficult to receive signals with a wide detection frequency band [13].

5. **Channelized receiver**

Channelized receivers were originally proposed to solve the problem of low-frequency accuracy of crystal video receivers. Placing multiple crystal video receivers in parallel narrows the sub-band, thereby improving the accuracy of frequency measurement. Since the number of crystal video receivers cannot be infinitely increased, the bandwidth is wide, and the frequency measurement requirements cannot be met, so the superheterodyne receivers are placed in parallel to obtain the basic model of the channelized receiver. Analog channelization receivers include SAW channelization receivers, bulk acoustic wave (BAW) channelization receivers and magneto-static wave (MSW) channelization receivers, depending on the analog components used in the filter. Taking the SAW channelized receiver as an example, the composition of a typical analog channelized receiver is shown in Fig. 6.9.

The signal processing of the analog channelized receiver is generally used for the detection of multiple channels, which is conducive to the discrimination of single-carrier frequency, frequency diversity, frequency coding (FSK), linear frequency modulation (LFM) and other signals. Its biggest feature is that the processing speed

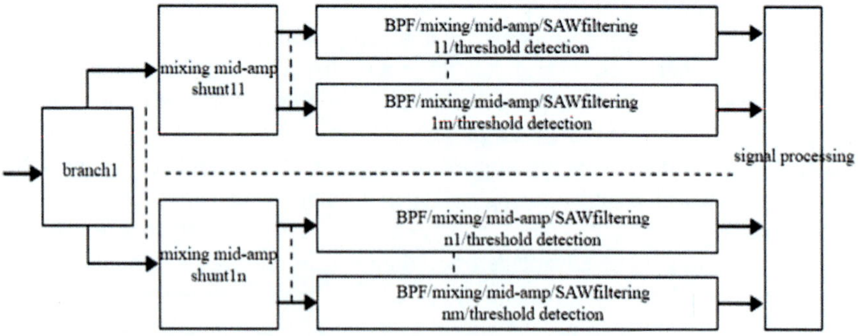

Fig. 6.9 Analog channelized receiver

is very fast, and it can better adapt to dense and complex multi-signal environments, but its equipment is usually relatively large.

The channelized receiver divides the received signal in the frequency domain, establishes the relationship between the frequency of the signal to be measured and the space power, thereby realizing the measurement of the frequency of the signal to be measured. The channelization receiver is mainly composed of the following parts: radio frequency front-end, power divider, filter and detector. The RF front-end is mainly responsible for low-noise amplification of the received signal. The basic idea is to parallelize a narrowband of a broadband RF signal. The power divider is responsible for dividing the low-noise-amplified signal into multiple channels, and then each filter uses a filter to filter out a part of the received RF signal, and the frequency domain division of the RF signal is achieved through the combination of the power divider and the filter. Then, each channel is detected by a detector, and finally, information such as the frequency of the received signal is judged according to the presence or absence of the detected power of different channels. Channelized receivers have many advantages, such as the wide frequency bandwidth of signals that can be detected by channelized receivers, which can overcome the shortcomings of instantaneous frequency measurement receivers that cannot handle signals that arrive at the same time, and achieve multiple RF signals that arrive at the same time. Real-time measurement to achieve full interception probability detection of the signal. The key component of a channelized receiver is a filter. In theory, the narrower the bandwidth of the filter, the greater the number of channels in the entire system, and the higher the frequency resolution of the entire receiver. At the same time, the narrower the bandwidth of the filter, the more the noise in each channel will be. The less the noise, the higher the sensitivity of the entire receiver. However, this type of channelized receiver is implemented based on analog devices. The high-frequency insertion loss of traditional analog devices is large, so the power consumption volume of channelized receivers based on traditional analog devices is very large, as the number of channels increases. The larger the volume power consumption of the entire system, the amplitude and phase between each channel are difficult to equalize. Therefore, in practical applications, the design indicators of channelized receivers must be comprehensively considered according to application requirements, system resources and technical difficulties. In addition, the channelization receiver has the problem of ambiguity in frequency measurement. In each channel, the detector directly detects each signal after passing through the filter, so the frequency resolution of the channelized receiver is the bandwidth of the filter for each channel, and it is difficult to accurately measure the frequency of the signal. At the same time, due to the spread spectrum and "rabbit ear" effects, it is necessary to judge the detection results between adjacent channels, thereby increasing the difficulty of designing the entire system and the complexity of the equipment.

Traditional electronic warfare receivers are based on traditional analog devices, and it is difficult to face the high frequency, broadband and dense signal environment of modern electronic warfare well. The output result of the crystal video receiver is rough, and the signals that arrive at the same time in the bandwidth cannot be resolved, and the interception probability is low. The superheterodyne receiver can measure high-frequency signals through multi-level down-conversion, but the multi-level down-conversion greatly increases with the complexity of the receiver, the volume of power consumption will increase significantly. At the same time, this type of receiver faces the contradiction between frequency resolution and scan time. Instantaneous frequency measurement receivers also face the disadvantage that they cannot measure the frequency of multiple signals that arrive at the same time. The above three receivers can be used in a modern war environment with high signal density. Compared with the previous three receivers, channelized receivers have obvious advantages. A channelized receiver can detect and receive multiple signals that arrive at the same time in a large frequency band. A channelized receiver is a receiver with a high probability of interception. Therefore, analog channelized receivers have been used in modern war environments. However, RF receivers based on analog devices have inherent disadvantages. It is difficult for analog electronic devices to achieve uniform high performance in one or several octave bands, resulting in low dynamic range, which makes traditional RF frequency measurement systems based on traditional microwave devices face the contradiction between frequency intercept probability and frequency resolution. In order to obtain a sufficiently high-frequency resolution, it is necessary to increase the number of sampling paths, which in turn leads to an increase in the SWaP of the device. At the same time, the microwave link based on the technology microwave device faces a large high-frequency loss, thereby limiting the frequency measurement fundamentally the measurement bandwidth of the system.

6.2.2 *Digital Channelized Receiver*

Facing a modern and complicated war environment, traditional analog receivers are increasingly difficult to meet the requirements. With the rapid development of analog-to-digital converters and modern digital signal processing technology, the introduction of digital signal processing technology into modern electronic warfare systems has become a trend of development. The working principle of the receiver based on digital signal processing is fundamentally different from the traditional analog receiver. All the signal processing part of the traditional analog receiver is processed in the analog domain, while the digital receiver first converts the analog radio frequency signal received by the antenna into a digital signal through a high-speed analog-to-digital converter, and then uses the high-performance digital signal processing module that processes the converted digital signal, thereby realizing the reception and processing of the signal. Compared with traditional analog receivers, digital receivers have the following advantages

1. It can make full use of the advantages of digital signal processing to achieve flexible, reconfigurable and high-precision measurement of signals. Through

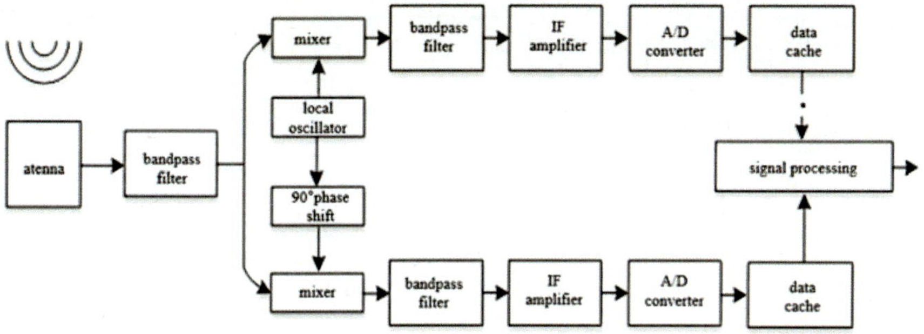

Fig. 6.10 Basic structure of a digital receiver

analysis of received signal pulses and between pulses, we can obtain characteristics such as the modulation format of the signal, which is more conducive to judging signal type.

2. The data can be stored for a long time through digital storage technology, and the signal can be delayed and processed in the digital domain. As long as the capacity of the digital storage device is large enough, the digital processing module can flexibly process the signals.
3. Digital signal processing will use I/Q demodulation and other technologies to ensure that the information of the received signal is completely retained and used for signal sorting and identification.

The basic structure of a digital receiver is shown in Fig. 6.10.

The digital receiver first processes the signal received from the antenna and then divides it into I and Q channels, each of which is mixed with the local oscillator, down-converts the signal to the intermediate frequency, and then filters out of band through a band-pass filter noise and nonlinear components from mixing. The signal after the IF amplifier is input to the analog-to-digital converter, the conversion of analog signals to digital signals is realized, and then the signal analysis and processing are realized through data buffering and digital signal processing. The core device of a digital receiver is an ADC for analog-to-digital conversion. With the increasing frequency of detection signals, ADCs that can meet both high speed and high accuracy have become the bottleneck of digital receiver development. Applications such as electronic countermeasures and frequency hopping communication require high-fidelity digital ultra-wideband signals and use DSP to extract target information from them. However, the increase in bandwidth makes the ADC quantization accuracy and power consumption details worse. In the face of massive bits (e.g., 12-bit quantization of a 5 GHz analog bandwidth will generate a bit rate of 60 Gb/s), the ADC will be powerless.

The digital channelization receiver performs AD conversion on the analog signal at the front-end and converts it into a digital signal. The digital filter bank is then used to replace the analog filter bank of the analog channelized receiver. A commonly used method for digital channelization of STFT digital channelization, the calculation

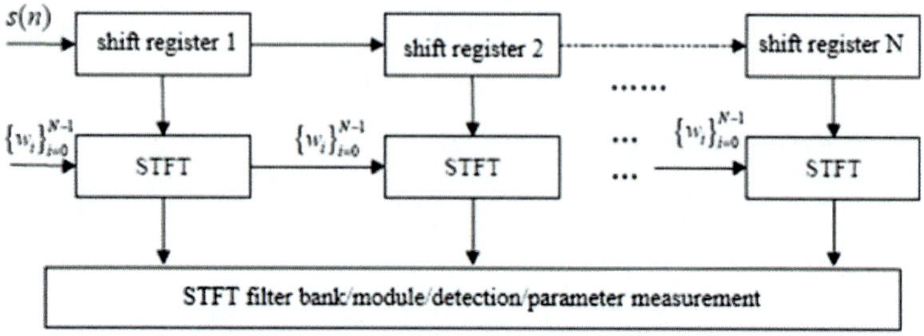

Fig. 6.11 STFT digital channelization

formula is

$$F(i,k)=\sum_{i=0}^{N-1}s(j+1)w_i W_N^{ik},\ k=0,\ldots,N-1. \tag{6.1}$$

The specific circuit structure implemented by Eq. (6.1) is shown in Fig. 6.11. Here, only one filter is drawn. If the circuit has sufficient resources, the circuit can implement point-wise continuous short-time Fourier transform operations.

The filtered processing flow is as follows:

(1) First perform modulo on the signal
(2) Then detect the signal to determine whether the threshold is crossed
(3) Then measure the specific parameters of the signal
(4) Count the power of each sub-channel when there is no signal output
(5) The noise power of each sub-channel can be obtained, and the constant false alarm detection of each sub-channel is performed
(6) Combining the detection results of each channel threshold, we can obtain the power spectrum of the wideband signal.

In order to solve the problem of processing speed, the above algorithm is improved. Let n = mp and make the following changes to Eq. (6.1):

$$\begin{aligned} F(i,k) &= \sum_{q=0}^{m-1}\sum_{j=0}^{p-1} s(mj+q+i)w_{mj+q}W_n^{(mj+q)k} \\ &= \sum_{q=0}^{m-1}\sum_{j=0}^{p-1} s(mj+q+i)w_{mj+q}W_p^{jk}W_n^{qk} \\ &= \sum_{q=0}^{m-1} F'(q,c,i)W_n^{qc}W_m^{qz} \end{aligned} \tag{6.2}$$

$$k=c+zp;\ z=0,\ldots,m-1;\ c=0,\ldots,p-1 \tag{6.3}$$

$$F'(q, c, i) = \sum_{j=0}^{p-1} s(mj + q + i) w_{mj+q} W_p^{jc}. \tag{6.4}$$

The algorithm first extracts m signal from the input signal and then performs parallel p-point short-time Fourier windowing filtering on the extracted signal. Then perform parallel spectral analysis on the output of these parallel filters.

6.3 Broadband Channelized Frequency Measurement Based on Cognitive Technology

Cognitive technology-based wideband channelization frequency measurement is cognitive digital channelization.

6.3.1 Principles of Cognitive Digital Channelization

Channelized receivers can detect signals with a 100% probability over a wide frequency band, and are common receiver types in electronic warfare systems. According to the organizational structure and method flow of the cognitive reconnaissance system, the specific process of cognitive measurement of signal frequency domain parameters is as follows:

Step 1: The signal enters the traditional digital channelization processing module. The traditional digital channelization channels are evenly distributed in the frequency domain (as shown in Fig. 6.12). A set of initial radar frequency domain parameters Fr0 is measured and the current digital channelization is recorded. Parameter setting Fj0;

Step 2: Fr0 and Fj0 are matched with the parameters in the knowledge base. If the matching is successful, the digital channelized reference settings and possible parameters of the radar signal are output. If the matching fails, the parameters are adjusted according to the existing rules to optimize parameters and update the learning library;

Step 3: Reset the digital channelization parameters and repeat the operations from Step 1 to Step 3 until the optimal measurement result is obtained.

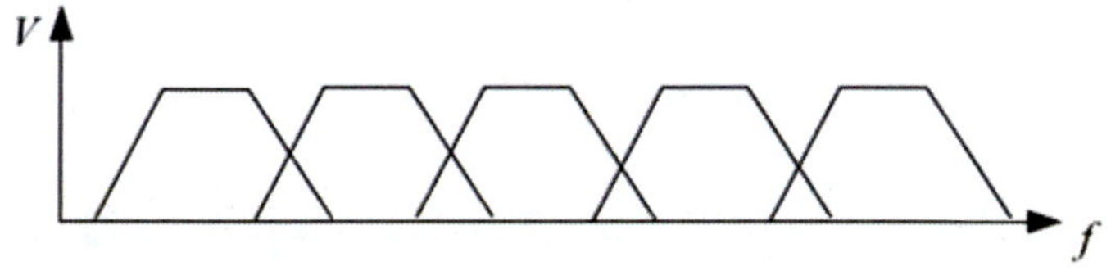

Fig. 6.12 Initial channel distribution

In Step 2, when changing the parameter settings of digital channelization, if the signal is a signal already in the cognitive knowledge base, read the channelization parameters directly in the cognitive knowledge base. For existing signals, the channelization parameters need to be changed. The detailed process is divided into single-channel parameter changes and multi-channel parameter changes based on whether the signals cross channels.

1. Multi-channel parameter change

When radar signals appear in multiple channels in the initial state, multi-channel parameter changes are used. For example, Fig. 6.13 shows the state of a signal with a bandwidth greater than the initial digital channel bandwidth after entering a digital channelization receiver. The red dotted line in the figure is the spectrum range of the input signal. As can be seen from Fig. 6.13, the signal spans the two channels of the digital channelized receiver. When the traditional method is used for processing, the signals in the two channels are considered as one signal, the two channels are processed separately, and the processing results perform analysis and identification to get the final result.

When frequency domain measurement is performed using digital channelization based on cognitive technology, the system dynamically adjusts the channel according to the signal parameters in each channel measured in Fig. 6.14. As shown in Fig. 6.15, during cognitive processing, if the signal bandwidth is greater than the system's initial bandwidth, the system combines the two channels to form a new digital channel; if the signal bandwidth is less than the system's initial bandwidth, the system moves the channel in the frequency domain to form a new digital channel.

2. Single-channel change

When the radar signal appears on a certain channel during the initial digital channelization (as shown in Fig. 6.16), a single-channel parameter change is used. The

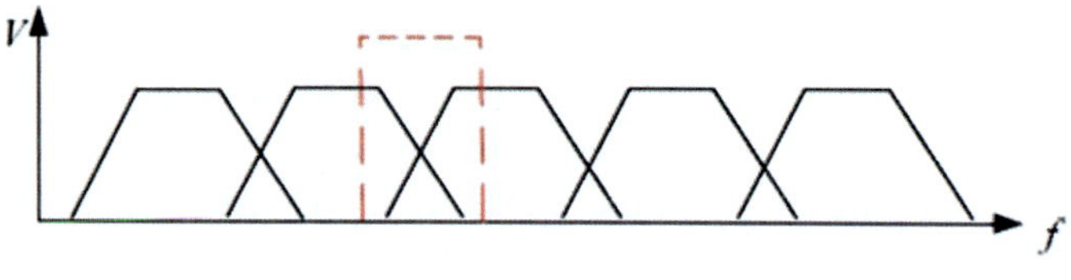

Fig. 6.13 Initial channel and signal of multi-channel parameter change

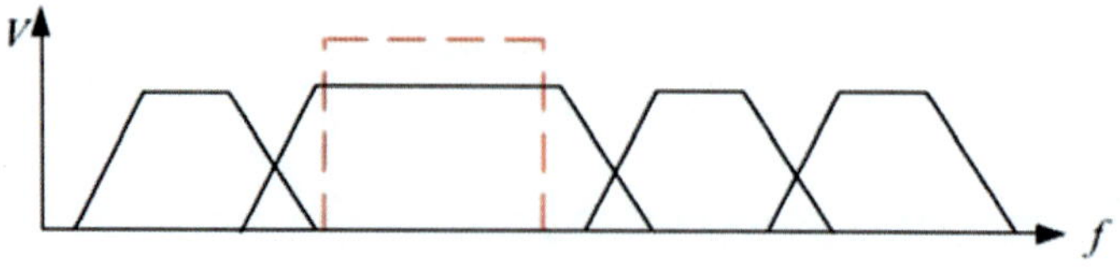

Fig. 6.14 Multi-channel parameter change cognitive processing synthetic channel

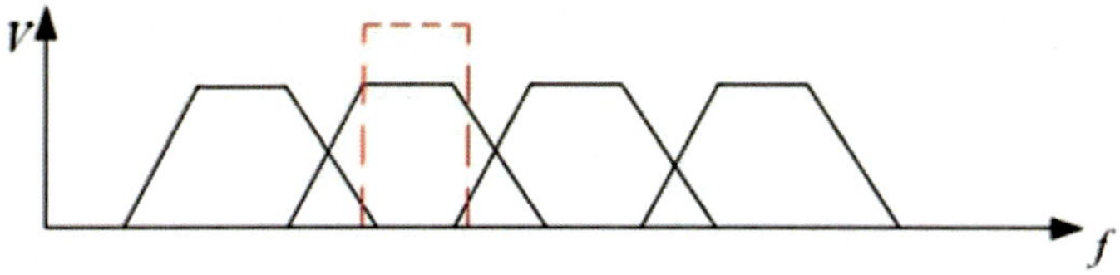

Fig. 6.15 Multi-channel parameter change cognitive processing moving channel

Fig. 6.16 Initial channel and signal of single-channel parameter change

red dotted line in Fig. 6.17 is the frequency spectrum of the input signal. This signal is a single-carrier frequency signal and appears on the second channel of the digital channelization receiver. The traditional method will produce an output on the second channel when processing and the accuracy of the frequency measurement are related to the FFT points of the channel.

When using digital channelization frequency domain measurement based on cognitive technology, the system dynamically adjusts the channel based on the signal parameters in each channel measured in Fig. 4.15. As shown in Fig. 4.16, after cognitive processing, the system divides a channel into two narrow channels and uses interpolation and other methods to increase the number of FFT points so that the number of FFT points in the two narrow channels is the same as the original channel, thereby further improving frequency measurement accuracy.

In cognitive digital channelization frequency measurement, the cognitive processing not only adjusts the parameters of the channel where the signal appears but also adjusts the parameters of the channel where the signal may appear. As shown in Figs. 6.18 and 6.19, the cognitive system searches and matches the cognitive knowledge base after analyzing the signal in the first channel where the signal appears and finds the fourth channel. If a signal is possible, the fourth channel is also processed to improve the response speed of the system.

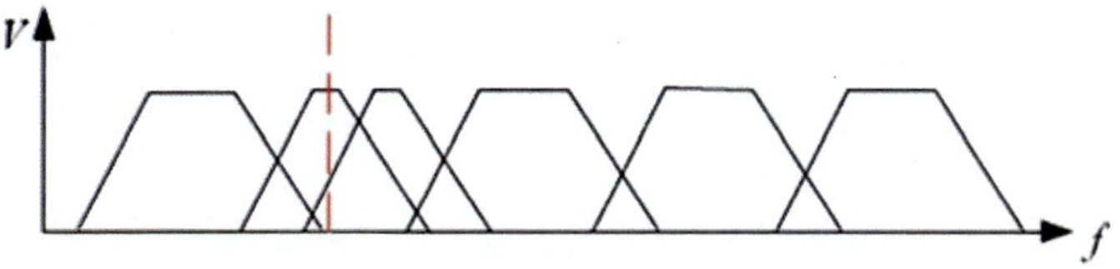

Fig. 6.17 Channel after single-channel parameter change cognitive processing

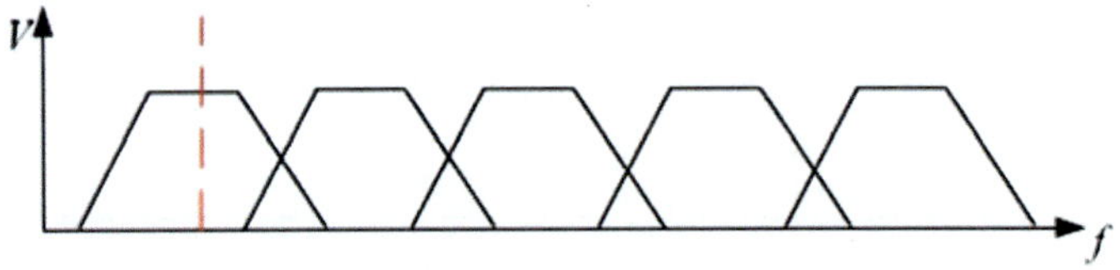

Fig. 6.18 Knowledge base search and processing of initial channels and signals

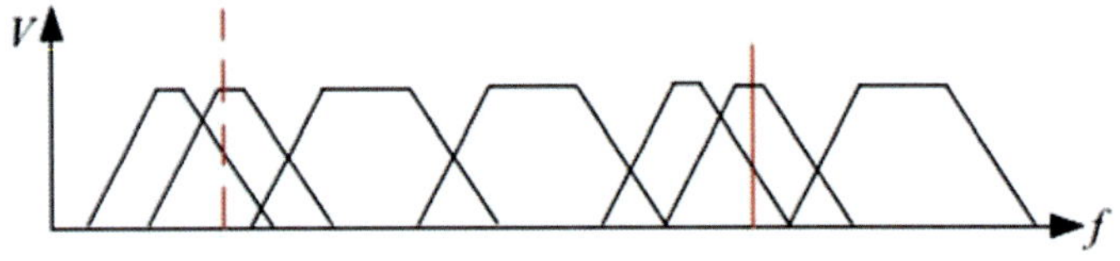

Fig. 6.19 Knowledge base search processing after cognitive processing channel

6.3.2 Cognitive Digital Channelization Features

DFT shows the frequency measurement accuracy of the digital frequency measurement method

$$\Delta f = \frac{B_f}{N}. \tag{6.5}$$

Among them, N is the number of points of the digitized signal, which is the range of non-fuzzy frequency measurement. In digital channelization frequency measurement, the accuracy of frequency measurement is fixed after the channel width and sampling frequency are fixed. In the cognitive digital system proposed in this section, the channel of the system adaptively adjusts the channel width after sensing the signal. In Fig. 6.18, with the same number of sampling points, reducing the channel bandwidth can significantly improve the accuracy of frequency measurement.

In channelization, the noise inside the receiver is white noise, and the signal-to-noise ratio of the signal is

$$SNR = \frac{P_s}{n_0 B_f}, \tag{6.6}$$

where P_s is the signal power and n_0 is the noise spectral density.

In channelization, when the signal crosses the channel, as shown in Fig. 6.8, the signal-to-noise ratio at this time is

$$SNR = \frac{P_s}{2n_0 B_0}. \tag{6.7}$$

Among them, B_0 is the channel bandwidth, and the signal-to-noise ratio is relatively low.

After the cognitive system processes the channel position, the signal is located in a channel, as shown in Fig. 6.10. The signal-to-noise ratio at this time is

$$SNR = \frac{P_s}{n_0 B_0}. \tag{6.8}$$

It can be seen that compared to traditional digital channelization, cognitive digital channelization processing significantly improves the signal-to-noise ratio when processing cross-channel signals.

References

1. DARPA, Notice of Intent to Award Sole Source Contract: Behavioral Learning for Adaptive Electronic Warfare(BLADE) Phase 3 [R/OL]. (2014–2–19). https://www.fbo.gov/spg/ODA/DARPA/CMO/DARPA-SN-14-24/listing.html.
2. R.A. Poisel, Electronic warefare receivers and receiving systems; K. Chang, *RF and Microwave Wireless Systems*. (Wiley, New York, 2000)
3. B. Razavi, *RF Microelectronics*, 2nd edn. (Pearson Education Inc., New Jersey, 2012)
4. C. Nicolai, L. Raffaelli, R. Rosati, Broadband crystal video front end for radar warning receivers. IEE Proc. F-Commun., Radar Signal Process. **129**(3), 133–137 (1982)
5. J.P.Y. Lee, Detection of complex and simultaneous signals using an instantaneous frequency measurement receiver. IEE Proc. F- Commun., Radar Signal Process. **132**(4), 267–274 (1985)
6. S. Foucart, H. Rauhut, *A Mathematical Introduction to Compressive Sensing* (Springer Science+Business Media, New York, 2013)
7. D.E. Allen, Channelised receiver: a viable solution for EW and ESM system. IEE Proc. F-Commun., Radar Signal Process. **129**(3), 172–179 (1982)
8. D.B. Chester, D.H. Damerow, C. Olmstead, Analog to digital converter requirements and implementations for narrowband channelization application, in *Acoustics, Speech, and Signal Processing* (IEEE, San Francisco, 1992), pp. 325–328
9. A.J. Seeds, K.J. Williams, Microwave photonics. J. Lightwave Technol. **24**(12), 4628–4641 (2006)
10. J. Capmany, D. Novak, Microwave photonics combines two worlds. Nat. Photonics **1**(6), 319–330 (2007)
11. J. Yao, Microwave photonics. J. Lightwave Technol. **27**(3), 314–335 (2009)
12. D. Marpaung, C. Roeloffzen, R. Heideman et al., Integrated microwave photonics. Laser Photonics Rev. **32**(16), 2849–2861 (2014)
13. I.P. Alcock, A.C. Tropper, A.I. Ferguson et al., Q-switched operation of a neodymium-doped monomode fibre laser. Electron. Lett. **22**(2), 84–85 (1986)

Chapter 7
The Basis of RF Photonic Channelization Technology

With the increase of communication bandwidth, the RF photon channelization receiving technology is to modulate broadband RF signals onto optical carriers, making full use of the broadband low-loss characteristics of optoelectronic devices, avoiding electronic bottlenecks and mapping the frequency and power of broadband RF signals to light. On the parameters of the carrier power, wavelength, time and space, the optical method is used to narrow the signal to realize the real-time reception of the sub-channel and sub-band of the radio frequency signal in parallel. The optical channelization receiving system is mainly divided into three parts: electro-optic modulation, spectral separation and photoelectric conversion, so the optical link is the basic structure of the receiver, which determines the overall performance of the receiver system. This chapter first introduces the structural characteristics and basic principles of the key components of each functional module of the channelized optical link. It analyzes the main indicators for evaluating the performance of the channelized receiving system, including link gain, noise, dynamic range and adjacent channel crosstalk. Structural characteristics of coherent detection are also analyzed. Then the performance of key dispersion devices such as dispersive fiber and Bragg grating used to achieve spectral separation is introduced. Finally, the photon under-sampling used to achieve down-conversion of multi-band RF signals is theoretically introduced. The content of this chapter is intended to provide a theoretical basis and support for subsequent chapters.

7.1 Basic Theory of Channelized Optical Links

7.1.1 Functional Structure of an Optical Link

If the analog optical link is regarded as a two-port network with RF input and output, it is an important mission of the optical link to realize high-performance transmission and processing of one-way or multi-directional radio frequency signals. A

J. Zhang and J. Li, *Satellite Photoelectric Sensing Technology*,
https://doi.org/10.1007/978-3-030-89843-4_7

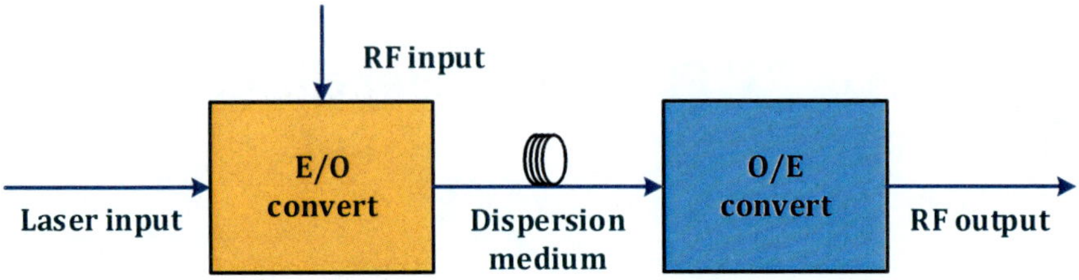

Fig. 7.1 Architecture of a typical channelized analog optical link

typical channelized analog optical link is mainly divided into three parts in terms of functional processing. The electro-optic modulation module realizes the mapping of microwave to light waves. The spectral separation part mainly uses dispersion devices to introduce different delays or diffraction angles to different spectra at different time or angle output; the photoelectric conversion module realizes light to electricity conversion and RF down-conversion function, as shown in Fig. 7.1. The core components of the entire radio frequency transmission are the light source, the electro-optic modulator and the photodetector. This section first gives a brief introduction to the main optoelectronic devices.

(1) Light source

As a carrier of radio frequency signals, a high-quality light source is a key to achieving high-resolution information perception in optical signal processing. It is the core device of the entire channelized optical link. According to the principle of the channelization technical solution, the system can selectively use continuous light sources and pulsed light sources, 1.31 and 1.55-micron wavelength light have the widest application range due to the zero-dispersion and lowest loss window of the fiber. Among various laser types, semiconductor lasers can generate light sources with adjustable wavelengths while maintaining stable dynamic single-vertical modeling characteristics by selecting different optoelectronic materials. The output optical power can be changed by adjusting the supply current, and it has a sufficiently narrow spectral width. It has become the primary choice for communication light sources with the advantages of simple modulation, high photoelectric conversion efficiency, small size, lightweight, low power consumption and high reliability. A distributed feedback (DFB) laser is shown in Fig. 7.2a for its structural characteristics and generation mechanism. DFB lasers have good monochromaticity, side touch suppression ratio and very small relative intensity noise [1]. The existing line width can reach the order of kHz. For the channelization scheme that uses pulsed light sampling to achieve signal down-conversion, ultra-short pulse sources with high repetition frequency, high peak power and wide spectrum are required. Mode-locked fiber lasers are the most widely used. The structure principle is shown in Fig. 7.2b, and Ti:sapphire laser, the pulse width has reached the order of femtoseconds.

As an information carrier of the entire optical link, the noise of the laser also has a large impact on the performance of the system, mainly including relative intensity noise (RIN) and phase noise. For ordinary intensity modulation direct detection

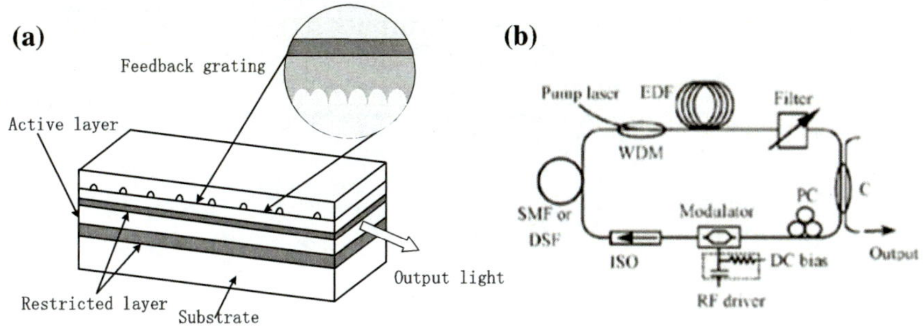

Fig. 7.2 Principle of a DFB laser and an active mode-locked fiber laser. **a** Structural characteristics of DFB lasers. **b** Principle device of active mode-locked fiber laser

systems, the influence of RIN noise plays a leading role. For a link using pulsed light, phase noise will cause the spectrum line of the pulse to widen and cannot be ignored [2, 3]. During the photoelectric conversion process of the RF signal down-converted from the optical domain to the electrical domain, the conversion of pulsed RIN noise to RF phase noise will also be generated, which will have a certain impact on the frequency stability of the RF signal [4, 5]. At the same time, for links with large dispersion media, there is also a conversion of phase noise to intensity noise, so in an actual link system, the relationship between laser noise and its performance on the link must be comprehensively considered according to the structure of the system impact.

(2) Electro-optic modulator

Electro-optic modulation is to map the information carried by the radio frequency signal to parameters such as the delay, amplitude, phase and polarization state of the optical carrier. According to the relationship between the light source and the modulation, it can be divided into direct modulation and indirect modulation. Direct modulation is to convert the modulation signal into a changed current to drive the laser to change the output optical power, which belongs to intensity modulation. However, the simultaneous occurrence of intensity modulation and the generation of light waves will cause the broadening of the spectral lines, which will be seriously affected by fiber dispersion during transmission, and the intensity noise of the laser will seriously deteriorate, so the bandwidth and modulation efficiency of the radio frequency signal that can be processed is limited. Indirect modulation is the separation of the laser and the modulator. The modulation device is used to implement the mapping of the microwave to the optical domain. According to different principles, it can be divided into different types, such as electro-optic effect, magneto-optical effect and electrical absorption effect. Indium phosphide (InP), lithium niobate (LiNbO3) and silicon-based modulators [6–9]. Lithium niobate modulators have become the most mature modulators with their wide adjustable bandwidth, high extinction ratio and high reliability.

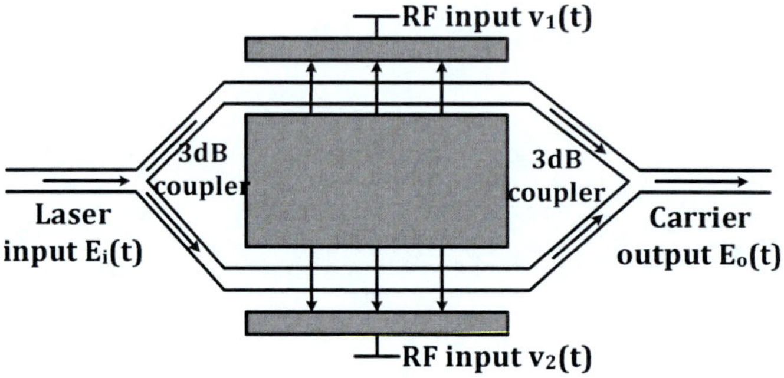

Fig. 7.3 Block diagram of a dual-drive Mach–Zehnder modulator

The most commonly used in channelized optical links are intensity modulation and phase modulation. Figure 7.3 shows a typical dual-drive Mach–Zehnder intensity modulator. The half-wave voltage is V_π, the input RF signal is expressed as $v_1(t)$ and $v_2(t)$, and the modulator's transmission function can be expressed as

$$S_{MZM} = \frac{1}{2}\left[1 + \cos(\Delta\phi) + \pi\frac{v_1(t) - v_2(t)}{V_\pi}\right], \tag{7.1}$$

where $\Delta\phi$ is the initial phase difference between the upper and lower arms of the modulator, which determines the working mode of the modulator, which can be adjusted by an external DC voltage. The formula (7.1) shows that the transmission function of the intensity modulator is a cosine function. By adjusting the DC bias, the intensity modulator can work in three typical working modes: low offset, quadrature offset and maximum transmission point. There is a high carrier rejection ratio at the low offset point, which can suppress the noise. The link gain at the orthogonal offset point is the largest and can suppress the even harmonics of the modulator. The highest offset point has maximum output power. In practical applications, different working modes can be selected according to different functional requirements. Phase modulation is the modulation of the RF signal onto the phase of the optical carrier, and its reception requires coherent detection. The specific introduction will be presented in the subsequent chapters.

(3) Photoelectric detector

The photoelectric converter is located at the output port of the optical link. It is the most critical module of the photoelectric system. It is the core device that realizes the modulation signal conversion from the optical domain to the electrical domain. Its working principle is based on the photoelectric conversion effect of the PN junction. At the intersection of P and N semiconductors, a self-built electric field is formed due to the difference in ion concentration [10]. Under certain conditions of photon incidence, electrons step up to form photo-generated electrons, and a drift current can be generated by applying an external electric field to the PN junction. A diffusion current is formed under thermal motion, and the sum of the two is the photo-generated current to achieve photoelectric conversion. Generally, there is a linear relationship between the photocurrent and the optical power. In this process, the performance of the photodetector device is reflected in whether it can efficiently and quickly convert photons to electrons, whether it can successfully detect weak beat signals and the ability to process high-power incident optical carriers. The influence of the internal electron parasitic devices and the diffusion speed of the carriers limit the response speed of the detector, which results in a limited detection wavelength range. Therefore, the operating bandwidth, conversion rate and responsiveness are the key indicators for measuring the performance of the detector. The detectors currently used are mainly divided into PIN photodiodes and avalanche photodiodes (APDs). Because APDs have gain effects, they can detect relatively low optical carriers, increase the link's gain while avoiding the use of subsequent electrical amplifiers. It is suitable for occasions with high receiving sensitivity requirements, but it will have nonlinear effect with the saturation of the input optical power. Commonly used are square photodetector (PD) and balanced photodetector (BPD). The balanced detector has two inputs and uses two photodiodes to make a difference between the current generated by the signal optical carrier and the local oscillator optical carrier. It can suppress the additive noise of the system. However, due to the use of two PDs, the shot noise introduced by the detector will double accordingly.

7.1.2 Performance Index of an Optical Link

The function of the photon channelized receiver is to realize the narrowband processing of wideband RF signals and to ensure the accurate reception of signals without distortion, distortion and omissions. As a two-port RF network composed of optoelectronic devices, the characteristics of optical devices directly determine the overall performance of the channelization system. This section details the performance evaluation of optical links based on the insertion loss, dispersion and nonlinear characteristics of each device.

(1) Link gain

Link gain (Gain) is the transmission and processing of the optical link. The power ratio before and after the RF signal input and output is the most basic performance parameter of the link. The specific formula can be expressed as

$$G = 10\log_{10}(P_{out} - P_{in}). \tag{7.2}$$

For a normal channelized optical link, modulation efficiency and photoelectric conversion efficiency are mainly affected by the link gain. The modulation efficiency is determined by the half-wave voltage and DC offset of the modulator. The photoelectric conversion efficiency depends on the response of the detector. For a simple electro-optical intensity modulation (MZM), photoelectric conversion optical network, the link gain can be expressed as

$$G(dB) = 20lg(I_d \frac{\pi R_s}{V_\pi} sin(\Delta\phi)). \quad (7.3)$$

Among them, $I_d = \alpha\gamma P_0/2$ is the detected photocurrent, α is the attenuation coefficient of the modulator, γ is the responsivity of the detector, P_0 is the optical power of the input detector, R_s is the matching impedance, V_π is the half-wave voltage of the modulator and $\Delta\phi$ is DC offset angle of the modulator. It can be seen from the above formula that the link gain is proportional to the photocurrent, and decreases with the increase of the half-wave voltage. When the offset angle is less than 90°, the link gain increases with the increase of the offset angle; when it is greater than 90°, it decreases with the increase of the offset angle; at 90°, it corresponds to the orthogonal offset point. At this time, the maximum link gain can be obtained. When the gain characteristics of the link are poor, the gain characteristics of the entire two-port network can be improved by adding front- and rear-stage microwave amplifier devices as a cascade system.

(2) Link noise

No matter in signal processing applications such as wireless communication or radar, noise will affect the transmission and reception quality of the signal, weaken the system's power processing range for the signal and reduce the sensitivity of the system. According to the direction of the signal in a typical optical link, the noise sources are divided into the relative intensity noise of the laser, the thermal noise of the resistive components, the spontaneous emission noise of the optical amplifier and the shot noise of the photodetector. Domain to electrical domain: understanding and studying the noise characteristics of each device and suppressing each noise to the minimum according to the application is of great significance to improve the performance of the link. Thermal noise exists in all electrical devices and is unavoidable. It is caused by the Brownian motion of the electrons. The direction of the current generated by the electrons is random and the average value is zero. It is only related to the bandwidth and ambient temperature.

$$P_{\text{thermal}} = kT\Delta f, \quad (7.4)$$

where k is the Boltzmann constant, T is the temperature and Δf is the noise bandwidth. At a room temperature of 25 °C and a noise bandwidth of 1 Hz, the thermal noise power is approximately −174 dBm. As long as the temperature is fixed, the thermal noise is constant and the upper limit of the noise performance of the link system. As the information carrier of the link, the noise characteristics of the laser have a great impact on the performance of the link. The noise introduced is mainly relative intensity noise:

$$P_{RIN} = \frac{I_d^2}{2} 10^{\frac{RIN}{10}} R_s \Delta f. \tag{7.5}$$

Among them, I_d is the detected photocurrent, R_s is the matching impedance and Δf is the noise bandwidth. In channelized receive optical links, RIN noise is a key factor. The relative intensity noise comes from the random radiation of photons and the random jitter of the pump current. It is the amplitude noise added to the output power, with an average value of 0. It is used to describe the stability of the laser output power. After photoelectric detection, the output caused by this part of the noise power has a linear relationship with RIN, which increases with the laser output power. The ideal optical amplifier is to achieve distortion-free amplification of light waves without introducing any noise. However, the mechanism of optical amplification will introduce spontaneous radiation noise to the signal, which is added to the output RF signal through photoelectric conversion. It is mainly related to optical power and amplifier gain. The spontaneous emission factor and the optical bandwidth are directly proportional. Shot noise is caused by the random fluctuations of photons and photo-generated carriers irradiated in the material during the process of converting photons into electrons by the detector. It belongs to the inherent noise of the detector. Noise is proportional to the magnitude of the photocurrent and noise bandwidth.

$$P_{\text{shot}} = 2eI_d R_s \Delta f, \tag{7.6}$$

where e is the charge constant. In general, different noise dominates under different input optical power conditions. The system behaves as shot noise limitation at lower powers and RIN noise limitation at higher input optical powers. Generally, the noise figure is used to characterize the noise characteristics of the entire link, which is used to measure the system's ability to handle weak signals:

$$NF = \frac{SNR_{\text{in}}}{SNR_{\text{out}}} = \frac{P_{N_{in}}}{G \cdot P_{N_{\text{out}}}}, \tag{7.7}$$

where SNR_{in} and SNR_{out} represent the input-to-output signal-to-noise ratio of the optical link, and $P_{N_{\text{in}}}$ and $P_{N_{\text{out}}}$ represent the system input and output noise, respectively. For an actual non-optically amplified receiving system, by measuring the noise floor of the link, the noise figure of the link can be calculated according to (5.8), where N is the measured total output noise of the link and G is the circuit gain. Thermal noise is − 174 dBm/Hz.

$$NF(dB) = N - G + 174. \tag{7.8}$$

(3) No spurious dynamic range

The spurious-free dynamic range (SFDR) refers to the input signal power range between the smallest signal that can be detected by the system and the largest signal that can be distinguished. It is the most critical performance indicator of the receiver system, and it is related to the noise and nonlinear characteristics of the link closely related. The noise characteristics have been introduced earlier, and the nonlinear characteristics of the link are first introduced below. Each component of the analog optical circuit, including the modulator, detector and microwave device, is not an ideal linear device. The optical link can be equivalent to a two-port nonlinear network, and the direct detection link (IMDD) can be modulated with a simple intensity. An example is used to analyze the nonlinear distortion of the link. For the MZM operating in the push-and-free mode, that is, the modulation signals of the upper and lower channels are the same size, the positive and negative are opposite, and the output from (7.1) can be obtained as

$$I(t) = I_d\left[1 + cos(\Delta\phi + \frac{\pi v_i(t)}{V_\pi})\right]. \tag{7.9}$$

When modulating a single-carrier signal $v_i(t) = Acos(\omega_{RF}t)$, the first kind of nth-order Bessel function expansion on (5.9) can be expressed as

$$I(t) = I_d\left\{\begin{array}{l} 1 + J_0(\beta)cos\Delta\phi \\ +2\cos\Delta\phi\sum_{n=1}^{\infty} J_{2n}(\beta)cos(2n\omega_{RF}t) \\ -2\sin\Delta\phi\sum_{n=0}^{\infty} J_{2n+1}(\beta)cos[(2n+1)\omega_{RF}t] \end{array}\right\}. \tag{7.10}$$

Among them, $\beta = \pi A/V_\pi$ is the modulation depth of the modulator, and $J_n(\beta)$ is the first-order nth-order Bessel function coefficient. As can be seen from the above formula, under the single-carrier input, in addition to the fundamental frequency signal, higher harmonic components also appear, which is the distortion introduced by the nonlinearity of the modulator. In the case of inputting a two-tone signal or multi-carrier, in addition to the fundamental frequency components and higher harmonics of each frequency component, intermodulation distortion and intermodulation distortion also appear. Usually even harmonics can be set by the orthogonality of the modulator. The offset point achieves greater suppression. The third and above harmonics can be filtered by filters because they are far away from the IF signal. As for the in-band intermodulation distortion and multi-carrier intermodulation distortion, linearization technology is required for suppression processing. The truncation point is an important parameter for objectively describing the magnitude of nonlinearity. It is the power value of the input signal when the power of the fundamental frequency signal is equal to the power of the nth intermodulation signal. The upper limit of the spurious-free dynamic range is determined by the nth truncation point. The link noise of the system is determined, as shown in Fig. 7.4.

(4) Channel balance and adjacent channel crosstalk

The equalization of the channel is used to describe the power jitter of the received signal, that is, the degree to which the shape of the frequency response of the channel deviates from the ideal rectangular window shape. Whether it is time-domain channelization or frequency-domain channelization, 100% interception of the signal must

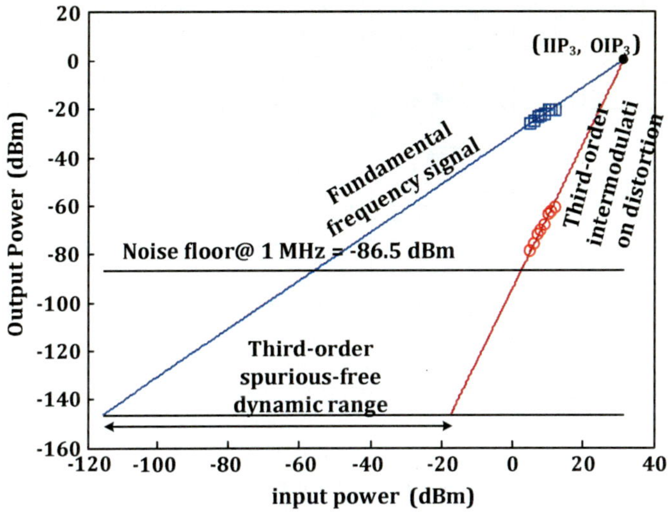

Fig. 7.4 Third-order spurious-free dynamic range

be guaranteed, and the frequency response. The unsatisfactory results will lead to the omission of the signal and the incomplete reception of the information, and the original signal cannot be accurately reconstructed and restored. The better the channel equilibrium, the closer the frequency response is to an ideal rectangular window. Adjacent channel crosstalk refers to the interference caused by the reception processing of adjacent or adjacent channel signals of a given channel. The channelized receiver divides the broadband RF signal into multiple narrowband frequency bands with the same bandwidth and sets multiple channels to seamlessly and parallelly receive RF signals of different frequency bands so as to realize broadband real-time perception reception. Multi-channel parallel work must ensure the channel the degree of isolation between the channels, avoids overlapping signals between channels, causing misjudgment of received information.

7.1.3 Coherent Reception Structural Characteristics

The conversion of optical signals to electrical signals by photoelectric systems can be divided into direct detection and coherent detection. Direct detection requires low coherence of the light source. The system is simple, easy to adjust and low in cost. However, it can only detect the average optical power of the input light limiting its application. Coherent detection is also called optical heterodyne detection, which uses a coherent light source. Unlike direct detection, it adds an optical local oscillator and performs a coherent beat frequency with the signal optical carrier. It can detect the amplitude, frequency and phase of the optical carrier. Parameters of which I/Q

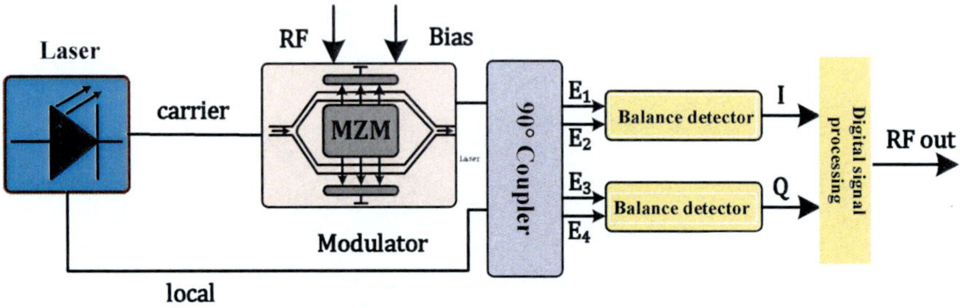

Fig. 7.5 Basic structure of I/Q coherent receiving system

demodulation combines the phase and amplitude information of the detection current can suppress the interference of the image frequency. The receiving performance is the best. The basic structure of I/Q demodulation is shown in Fig. 7.5. Coherent detection requires high coherence of the light source. To ensure that the frequency between the signal light and the local oscillator is relatively stable and interference can occur, the same laser is generally used in the system, and the output light is divided into two channels, one of which is an optical carrier. All the way it serves as the optical local oscillator, while the line width of the laser must be narrow enough to reduce the phase noise. The local oscillator can set a relatively large optical power, increase the output photocurrent and have a high conversion gain, which can enhance the perception of weak signals and thus improve the receiving sensitivity of the optical link. In addition, the direct receiving system usually uses square detection, which can only process signals with intensity modulation, while coherent detection can handle carriers with multiple modulation methods including intensity modulation, phase modulation and polarization modulation, which increases the flexibility of the system. With the increase of the working bandwidth of the detector and the performance improvement of the laser in the output power, amplitude and frequency stability, line width and other aspects, the application of coherent communication is more extensive.

In coherent optical communications, the power of the local oscillator light is usually relatively large, which will introduce higher relative intensity noise. The photoelectric detection is usually implemented by balanced detection. The balanced detection structure usually includes a 90° optical coupler and two balanced photodetectors. As shown in Fig. 7.5, the back-end uses digital signal processing to achieve signal recovery. This method can suppress the additive noise of the system including the RIN noise of the laser. Assuming that the link uses phase modulation, the input RF signals and the carrier and local oscillator signals can be expressed as

$$E_{sig}(t) = \sqrt{P_s}e^{i[\omega_0 t + \frac{\pi}{V_\pi} v_{in}(t) + \phi_1(t)]} \tag{7.11}$$

$$E_{lo}(t) = \sqrt{P_{lo}}e^{i[\omega_0 t + \phi_2(t)]}, \tag{7.12}$$

where $P_s = \alpha P_0(1 + n(t))$ and $P_{lo} = P_0(1 + n(t))$ are the optical power of the signal and the local oscillator, α is the loss coefficient of the phase modulator, n(t) is the relative intensity noise of the laser, and $\phi_1(t)$ and $\phi_2(t)$ are the phase noise of the two channels of light, respectively. To a 90° optocoupler with two inputs and four outputs, its transfer function and output can be expressed as

$$\begin{bmatrix} E_1 \\ E_2 \\ E_3 \\ E_4 \end{bmatrix} = \frac{1}{2}\begin{bmatrix} 1 & 1 \\ 1 & -1 \\ 1 & j \\ 1 & -j \end{bmatrix} \cdot \begin{bmatrix} E_{sig}(t) \\ E_{lo}(t) \end{bmatrix} = \frac{1}{2}\begin{bmatrix} E_{sig}(t) + E_{lo}(t) \\ E_{sig}(t) - E_{lo}(t) \\ E_{sig}(t) + jE_{lo}(t) \\ E_{sig}(t) - jE_{lo}(t) \end{bmatrix}. \tag{7.13}$$

The output of the four signals through two balanced detectors can be expressed as

$$I = E_1^2 - E_2^2 = \gamma\sqrt{P_s P_{lo}}\cos(\frac{\pi}{V_\pi}v_{in}(t) + \varphi(t)) \tag{7.14}$$

$$Q = E_3^2 - E_4^2 = \gamma\sqrt{P_s P_{lo}}sin(\frac{\pi}{V_\pi}v_{in}(t) + \varphi(t)), \tag{7.15}$$

where $\varphi(t) = \phi_1(t) - \phi_2(t)$. By digital signal processing on Eqs. (7.14) and (7.15), two outputs can be obtained:

$$W = I + jQ = \gamma\sqrt{P_s P_{lo}}e^{j(\frac{\pi}{V_\pi}v_{in}(t) + \varphi(t))}. \tag{7.16}$$

Finally, perform linear demodulation on (7.16) to recover the RF signal:

$$y(t) = Im[ln(W)] = \frac{\pi}{V_\pi}v_{in}(t) + \varphi(t). \tag{7.17}$$

From the analysis of (7.17), it can be seen that using a coherent receiving system combining balance detection and digital signal processing, compared with direct detection, the post-processing of the electrical signal is converted to the digital domain, and finally all the information of the modulated RF signal can be obtained. In addition, the use of some electrical components is avoided, so no matter what modulation method is used, the complexity of the system will not be increased. In addition, the fiber dispersion introduced in the link, the nonlinear distortion of the modulator and the polarization mode dispersion can be compensated and suppressed by algorithms in the digital processing in the later stages, thereby improving the dynamic range and sensitivity of the system. In summary, coherent reception will have great application value in the future communication field.

7.2 The Main Dispersion Device for Photon Channelization

The core functional module of photon channelized RF reception is the separation and cutting part of wide-spectrum light. Usually, the uneven distribution of refractive index in the dispersive device is used to introduce different group delays to different frequency wavelengths to achieve different reception times or diffraction angles. The effects of dispersion and nonlinearity of the device will have negative effects on signal transmission and processing such as pulse broadening and inter-symbol interference, which will seriously affect the transmission capacity and communication distance of information. Channelized reception is to take full advantage of this feature, to achieve spectral separation, achieve channel division from the optical domain, or introduce chirp by pulse broadening, and to use the characteristics of chirp pulses to perceive information from the time domain [11–13]. This section mainly introduces commonly used dispersion. Optical fibers and diffraction gratings are used for the structural and functional characteristics of spectral separation devices.

7.2.1 *Structural Characteristics of Dispersive Fiber*

As the main information transmission medium in an optical link, optical fiber is a waveguide structure with a circular cross-section. It is mainly made of quartz fiber. Combined with appropriate impurities, different refractive index distributions can be achieved. The propagation of light in it must meet the Maxwell equation and electromagnetic field boundary conditions. According to the different conduction modes, it can be divided into single-mode fiber and multi-mode fiber. The core diameter of single-mode fiber is relatively small, usually only 4–10 μm. It can only transmit one mode, which is suitable for large-capacity and long-distance communication. There are many different transmission modes, which are suitable for mid-range transmission. Transmission characteristics such as the loss of light, dispersion and nonlinear effects will directly affect the communication quality. In the optical link long-distance communication system, the loss of the optical fiber cannot be ignored, which determines the relay distance of the communication. Generally, the loss characteristic is measured by the power attenuation or loss coefficient introduced within the unit length. The main factors leading to fiber loss are the intrinsic absorption of quartz materials, Rayleigh scattering, structural defects introduced by imperfect manufacturing processes and leakage caused by pressure bending during the use of optical fibers. From the loss spectrum of the optical fiber, as shown in Fig. 7.6, it can be seen that the loss coefficient decreases with the increase of the wavelength as a whole. There are three windows with extremely low loss, which are 850, 1310 and 1550 nm band. The attenuation of the commonly used fiber at the 1550 nm window is reduced to 0.2 dB/km, close to the limit.

In addition to loss, dispersion is another important parameter that affects optical fiber communication. When an optical fiber transmits an optical carrier with a certain

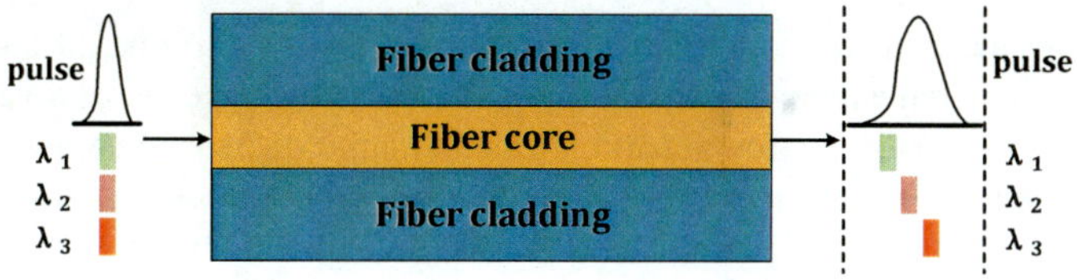

Fig. 7.6 Schematic diagram of the loss spectrum of a single-mode fiber

spectral width, the signal energy is composed of multiple frequency components, and different frequency components have different transmission rates, which results in dispersion phenomena such as Fig. 7.6. In single-mode fiber, fiber dispersion includes material dispersion, structural dispersion (collectively called chromatic dispersion, both of which coexist) and polarization mode dispersion. There are many different ways to describe fiber dispersion from different angles, such as dispersion coefficient, maximum delay difference and fiber bandwidth. The most widely used is the delay difference introduced by a unit line width in a unit length fiber, which is the dispersion coefficient. The characterization is given by

$$D(\lambda) = \frac{\Delta\tau(\lambda)}{\Delta\lambda}. \tag{7.18}$$

The dispersion coefficient $D(\lambda)$ is ps/nm km, $\Delta\tau(\lambda)$ is the delay difference introduced per unit length and $\Delta\lambda$ is the optical carrier line width. The chromatic dispersion of single-mode fiber is negative dispersion below 1310 nm. As the wavelength increases, the negative dispersion decreases. 1310 nm is the zero-dispersion window of the fiber. Then the positive dispersion gradually increases with the wavelength. The dispersion coefficient at 1550 nm is 17.4 ps/nm km. In general, the main factor that causes pulse broadening in an optical fiber is group velocity dispersion, that is, second-order dispersion, and the second-order dispersion coefficient can be expressed by the formula (7.19):

$$\beta_2 = \frac{d^2\beta}{d\omega^2}, \tag{7.19}$$

where β is the phase constant of the light wave. For optical fiber communication, dispersion seriously affects the signal transmission quality, so dispersion compensation is particularly important. The basic principle is to use a larger negative dispersion device to cancel the positive dispersion of the fiber and reduce the total dispersion. At present, the commonly used compensation technology is divided into fiber-type dispersion compensation and grating-type dispersion compensation [104]. The dispersion compensation fiber (DCF) adopts a multi-clad structure to obtain a large negative dispersion coefficient and a large negative dispersion slope. As a passive device, the installation is flexible, and the application is now mature. Dispersion compensation gratings compensate for the delay difference introduced by dispersion by changing the transmission paths of different light waves so that all the spectral components reach the receiving end at the same time to achieve dispersion compensation. With its small size, low insertion loss and large dispersion compensation, it is widely used.

For link systems with high optical power and low optical fiber loss, the nonlinear effects of optical fibers cannot be underestimated. They are mainly divided into two categories: stimulated scattering effects and refractive index perturbations. The stimulated scattering effect is divided into two types: stimulated Brillouin scattering and stimulated Raman scattering. In the case of large incident light power, the stimulated scattering effect can be regarded as a loss characteristic and increased exponentially. In addition, as the optical power increases, the refractive index distribution of the fiber changes accordingly, which results in refractive index perturbations. The refractive index perturbations will cause self-phase modulation (SPM), cross-phase modulation (XPM), four-wave mixing (FWM) and other nonlinear effects [90]. Self-phase modulation is an additional change of the intensity of a light wave to its own phase over time. The changing phase, that is, the frequency of the pulse, will cause the pulse to broaden. Cross-phase modulation is a phenomenon in which the phase of a certain wavelength of light is modulated by the intensity changes of other wavelengths. It is used to describe the interaction between different wavelengths. In photon channelization, it means that different channels will crosstalk with each other. The four-wave mixing effect is a process in which two or three different wavelengths of light waves interact with each other to generate two or three new optical frequencies under certain phase conditions. It belongs to intermodulation distortion. For channels that use signal multicast technology, four-wave mixing can also cause crosstalk between channels. In summary, both the dispersion and nonlinear effects of optical fibers will affect the channelized reception, and the characteristics of optical fibers must be considered in application systems of different schemes.

7.2.2 Fiber Bragg Grating

Fiber Bragg grating (FBG) is a passive optical device developed based on the sensitivity of optical fiber. When ultraviolet light is written, the refractive index of the fiber core will change periodically with the spatial distribution of light intensity That is, the refractive index is modulated to generate a permanent diffraction grating. A Bragg grating is essentially a narrowband periodic mirror or filter. As shown in the cascade weak reflection (CWRM) model in Fig. 5.7, each period is equivalent to one mirror, and FBG can be regarded as multiple. The mirrors are arranged at equal intervals. When broad-spectrum light is incident, light that meets the Bragg condition (7.20) is reflected, and the rest of the spectrum is transmitted:

$$\lambda = 2n_{\mathrm{eff}}\Lambda/m, \tag{7.20}$$

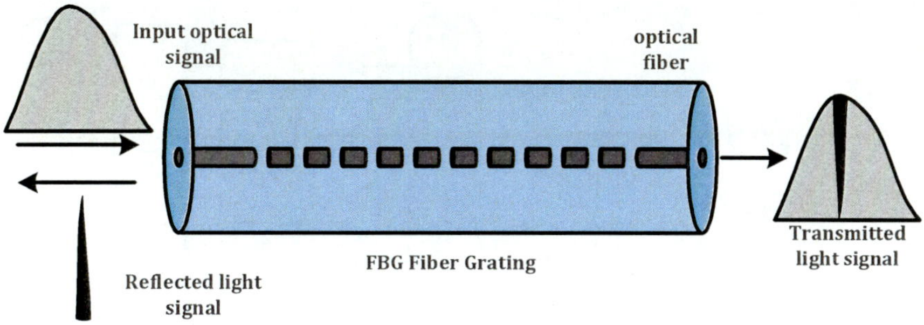

Fig. 7.7 Schematic diagram of the fiber Bragg grating

where λ is the incident wavelength, n_{eff} is the effective refractive index, Λ is the period of the grating and $m \geq 1$. With $m = 1$, $n_{\text{eff}} = 1.455$ and an incident wavelength of 1.55 μm, the period of the grating is 532.65 nm. The relationship between the period and the reflectivity described in the above formula is similar to the Fourier transform. According to the periodic classification, fiber gratings can be divided into short-period fiber gratings and long-period fiber gratings (LPFG). The propagation of light waves in a core with periodic dielectric disturbances will cause coupling between different modes, resulting in energy exchange. The coupling mode theory (CMT) of optical waveguides is the main method to describe the performance of FBG. The dielectric is discontinued on a periodic basis [14–17]. The Fresnel formula is used to describe the transmission characteristics of light waves. The perturbation of the refractive index obtained through approximate processing can be expressed as a modulated sine wave with additional DC components. Different modulation methods, such as apodization and apodization, can get gratings with different functions.

With the development of wavelength division multiplexing technology, the function of FBG has evolved from simple narrowband optical filtering to multi-channel spectral filtering and dispersion compensation between different wavelengths in wideband RF channelization reception [18]. FBG has strong control over the group delay of light, and the grating technology with a complex dispersion compensation function is very mature. For example, tunable second-order group delay FBG and third-order group delay FBG have been put into practical use [19]. Fiber gratings are widely used in communication, sensing and biochemical fields due to their simple and flexible UV engraving methods, strong anti-interference ability, small size and easy reuse. The schematic diagram of the fiber Bragg grating is shown in Fig. 7.7.

7.3 Optical Sampling Link Based on Pulse Source

With the improvement of communication capacity, communication speed and hardware level, digital signal processing has become the mainstream trend of signal processing with its characteristics such as great flexibility, high accuracy, low

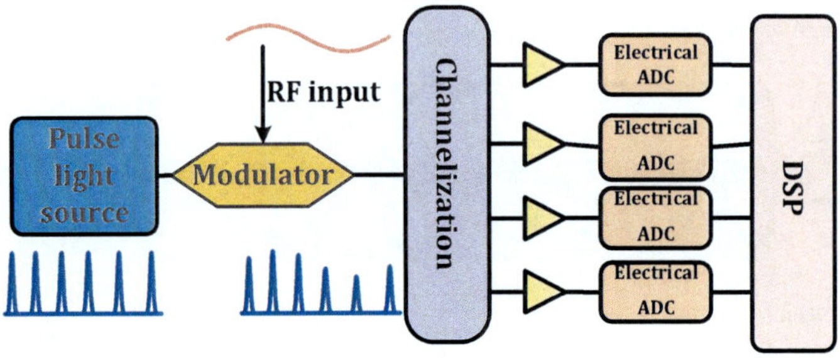

Fig. 7.8 Structure of channelized reception based on optical sampling

power consumption and easy integration. As a key functional module that converts analog signals into digital signals, analog-to-digital conversion gradually transitions from traditional electrical ADCs to optical ADCs. Optical analog-to-digital conversion relies on the high-speed, broadband and low-loss characteristics of light to break through the bottleneck of electrical analog-to-digital conversion [20–22]. It provides a high-speed and effective technical means for the realization of analog-to-digital conversion. Analog-to-digital conversion is mainly divided into two parts: sampling and quantization. Optical ADC does not mean all-optical processing, and an electrical-assisted solution can also be used. According to the role of light in analog-to-digital conversion, optical ADCs can be divided into four categories: optical assistance, optical sampling, optical quantization and all-optical ADCs. The more mature applications are optical conversion technologies that use optical sampling as an aid. The channelized optical analog-to-digital conversion link uses pulsed light to sample radio frequency signals. The back-end electronic ADC assists the quantization system structure, as shown in Fig. 7.8. This section mainly details the basic principles of the optical sampling link.

The sampling theorem is a bridge for converting analog signals into digital signals and is an important basis for the discretization of continuous signals. In simple terms, when the sampling frequency is greater than twice the maximum frequency of the signal, the signal can be completely recovered from the sampled samples without spectral mixing. Overlapping the dispersion in the time domain corresponds to the periodic extension in the frequency domain, as shown in Fig. 7.9. In the optical sampling system, a pulsed light source with a narrow line width and a small time jitter is used as the sampling pulse. The line width of the laser determines the time resolution of the sampling. The time jitter of the pulse affects the time jitter of the entire system. In practice, most signals are bandpass, and the signal bandwidth is far less than the maximum frequency. The sampling frequency does not need to meet the Nyquist sampling frequency. As long as the bandpass sampling theorem is used, the signal can be reconstructed without distortion, which reduces subsequent processing speed requirements.

For a band-pass signal whose frequency component belongs to $[f_l, f_u]$ and the signal bandwidth is $B = f_u - f_l$, to ensure that no spectral aliasing occurs, the sampling rate f_s needs to satisfy

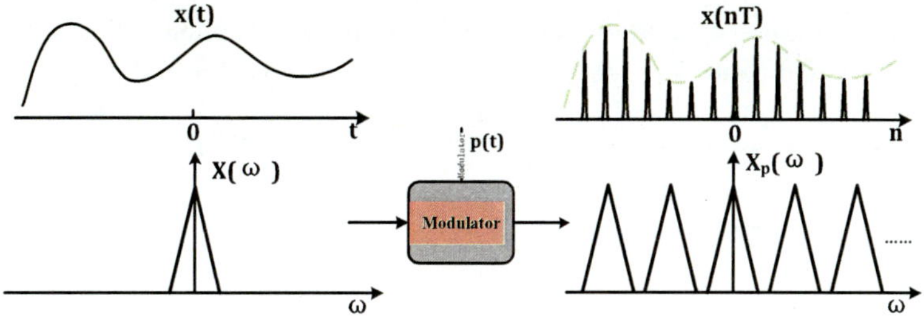

Fig. 7.9 Schematic diagram of spectrum shifting achieved by sampling

$$\frac{2f_u}{n} \le f_s \le \frac{2f_l}{n-1}, \qquad 1 \le n \le N. \tag{7.21}$$

Assuming N is the largest positive integer not greater than f_u/B, and n is an integer, the spectrum can be completely transferred to the first Nyquist bandwidth under the condition of formula (7.21). The baseband signal is filtered out. However, during the process of spectrum shift, noise will also be superimposed, resulting in deterioration of the signal-to-noise ratio. The condition for the equal sign in the above formula is that the band-pass signal has no frequency components at f_l and f_u. Figure 7.10a intuitively describes the relationship between the sampling rate and signal bandwidth expressed in (7.21) [23, 24]. The ordinate is the sampling rate normalized by the channel bandwidth, the abscissa is the position of the passband signal and the shaded part indicates the sampling rate interval that will cause spectral aliasing. When n = 1, there is $f_s \ge 2f_u$, which corresponds to the Nyquist sampling theorem. In theory, the minimum sampling rate corresponding to each sampling rate interval is $f_s = 2B$, which is located at an integral multiple of the highest frequency of the signal bandwidth. However, in practical applications, any minor imperfections in system engineering will cause the spectrum to drop. It can also be seen from the figure that the selection of the sampling frequency is not as high as possible, and there

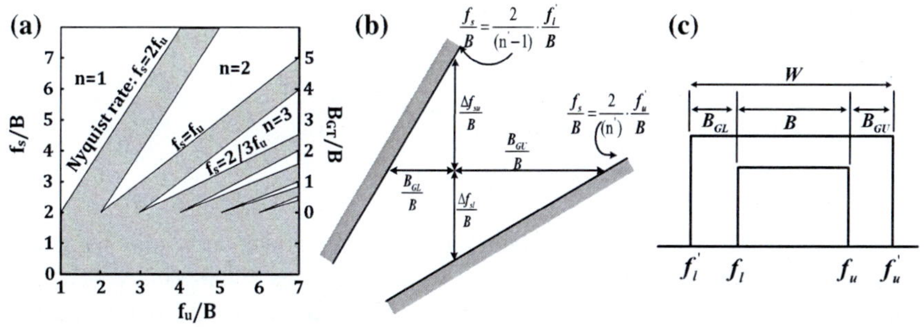

Fig. 7.10 The relationship between the sampling rate and sampling bandwidth of the band-pass sampling theorem (**a**). The selection interval of the sampling rate without distortion (**b**). Boundary conditions for distortion-free sampling rate intervals (**c**)

are upper and lower limits. Sampling at a non-minimum sampling rate is equivalent to adding a guard band B_{GT}, $B_{GT} = B_{GL} + B_{GU}$ to avoid spectral aliasing to the signal bandwidth, and $B_{GT} = f_s - 2B$, the signal bandwidth is expanded to W, and the frequency interval is expressed as $\left[f_l', f_u'\right]$, where $f_l' = f_l - B_{GL}$, $f_u' = f_u + B_{GU}$, as shown in Fig. 5. As shown in Fig. 7.10b, the selection range Δf_s of the sampling rate can be expressed as

$$\Delta f_s = \Delta f_{su} + \Delta f_{sl} = \frac{2f_1'}{n' - 1} - \frac{2f_u'}{n'}, \tag{7.22}$$

where n' is the largest integer not greater than f_u'/W. If the sampling rate is at the midpoint of an interval perpendicular to the ordinate, then

$$f_s = \frac{1}{2}\left(\frac{2f_u'}{n'} + \frac{2f_l'}{n' - 1}\right). \tag{7.23}$$

According to formula (7.23), the upper and lower boundary conditions of the sampling interval can be obtained, as shown in Fig. 7.10c. Generally, the addition of a guard band is equivalent to increasing the tolerance of signal distortion, and it can avoid subsequent filters. The rising and falling edges of the filter eliminate signal loss caused by unwanted filters.

In the process of moving the signal to the baseband by the band-pass optical sampling, the noise of multiple passbands also moves with the signal. Assume that the power spectral density of the signal is S, the in-band noise power spectral density N_p, and the out-of-band noise power spectral density N_0. Noise aliasing will lead to a reduction in the signal-to-noise ratio SNR_s:

$$SNR_s \approx \frac{S}{N_p + (n - 1)N_0} \tag{7.24}$$

If the in-band noise is much greater than the out-of-band noise, the signal-to-noise ratio is determined before sampling. If the in-band noise is the same as the out-of-band noise, in the case of $n \gg 1$, the deterioration of the signal-to-noise ratio can be expressed as

$$D_{SNR}(dB) \approx 10 \log n \tag{7.25}$$

The above formula is obtained on the premise of flat power spectral density and ideal low-pass filtering. For example, for a signal with a bandwidth of 30 kHz, the highest frequency is 455 kHz, n is 30, or the highest frequency is 10.7 MHz, and n is 356. The noise power increases by at least 15 and 25 dB. When the noise spectrum is not flat, the average power spectral density is used to measure. It is assumed that the equivalent average power spectral density of the analog signal noise before sampling is N_{EA}, the noise bandwidth is B_{EA} and the equivalent average power spectral density of the noise after sampling is NES. If the bandwidth is B, then they meet

$$N_{EA}B_{EA} = N_{ES}B. \tag{7.26}$$

Then the degradation of the signal-to-noise ratio N_{ES}/N_{EA} can be expressed as

$$D_{SNR}(dB) \approx 10\log\left(\frac{B_{EA}}{B}\right). \tag{7.27}$$

For a channelized receiving system, the noise level of adjacent channels may be much higher than the in-band noise, so the signal-to-noise ratio deteriorates more severely. For most communication systems, in order to improve the signal-to-noise ratio, different filters are usually selected for anti-spectral aliasing according to performance requirements. For sampling systems with anti-aliasing filters, the signal degradation ratio (SDR) is the ratio between the mean square error of the noise passing through the filter and the total energy of the signal. An important basis for different filters is

$$SDR = \frac{\int_0^{\infty} |H(\omega)|^2 d\omega}{\int_0^{2\pi f_c-(\pi/2)f_s} |H(\omega)|^2 d\omega + \int_{2\pi f_c+(\pi/2)f_s}^{\infty} |H(\omega)|^2 d\omega}, \tag{7.28}$$

where $H(\omega)$ is the transfer function of the filter and is the center frequency of the filter. The above formula only provides an improved way to reduce signal distortion. In addition, the signal-to-noise ratio is also affected by many other factors. For example, the final signal quality is also closely related to the performance of the laser. The quality of the sampling pulse is critical to the final quantization accuracy. The line width, amplitude jitter and time jitter of the pulse and the effective digits of the quantization satisfy the following relationship [25, 26]:

$$\tau_e \leq \frac{1}{\pi f_s}\left(\frac{12}{2^{ENOB-1}}\right)^{1/2} \tag{7.29}$$

$$\tau_a \leq \frac{1}{\sqrt{12}\cdot 2^{ENOB-1}} \tag{7.30}$$

$$ENOB = \log_2\left(\frac{1}{\sqrt{3}\pi f_s\sigma}\right), \tag{7.31}$$

where τ_e is the full-width at half-maximum of the pulse, τ_a is the amplitude jitter of the pulse and σ is the time jitter of the laser pulse. With the development and maturity of high-performance mode-locked lasers such as high repetition rates and narrow line widths, the application prospects of optical sampling technology are more extensive.

7.4 Summary

Starting from the basic structure of a typical channelized analog optical link, this chapter introduces the main functions of core components including light sources, electro-optic modulators and photodetectors, from the link's gain, noise figure, dynamic range and adjacent channel crosstalk. The performance index of the channelization system is analyzed in other aspects, and a mathematical theoretical model of coherent reception and linear demodulation is given in combination with the balanced detection technology. Secondly, the common methods of channelized spectral separation are discussed, and the structural principles and performance parameters of dispersive fiber and Bragg grating are analyzed. Finally, a channelized receiving system model based on optical sampling is established. The selection rules of alias-free sampling rate in band-pass sampling are discussed in detail. The influence of band-pass sampling on the signal-to-noise ratio and the line width, time jitter and amplitude of the laser are analyzed. There is a correlation between jitter and subsequent quantized significant digits. This chapter aims to provide the necessary theoretical basis for subsequent chapters.

References

1. D.C. Schleher, Electronic warfare in the information age. Artech House Radar Library
2. K. Wu, C. Ouyang, J.H. Wong et al., Frequency response of the noise conversion from relative intensity noise to phase noise in the photodetection of an optical pulse train. IEEE Photonics Technol. Lett. **23**(8), 468–470 (2011)
3. R. Paschotta, Noise of mode-locked lasers (Part II): timing jitter and other fluctuations. Appl. Phys. B **79**(2), 163–173 (2004)
4. S. Yamamoto, N. Edagawa, H. Taga et al., Analysis of laser phase noise to intensity noise conversion by chromatic dispersion in intensity modulation and direct detection optical-fiber transmission. J. Lightwave Technol. **8**(11), 1716–1722 (1990)
5. W.K. Marshall, B. Crosignani, A. Yariv, Laser phase noise to intensity noise conversion by lowest-order group-velocity dispersion in optical fiber: exact theory. Optical Lett. **25**(3), 165–167 (2000)
6. M.R. Salehi, B. Cabon, Y. Le Guennec, Influence of the chirp effect of DFB laser in phase-to-intensity noise conversion in RF-modulated optical links, in *MTT-S International Microwave Symposium Digest* (IEEE, Philadelphia, 2003), pp. 1371–1374
7. G.W. Anderson, Advanced channelization technology for RF, microwave and millimeterwave applications. Proc. IEEE **79**(3), 355-388P (1991)
8. S.T. Winnall, A.C. Lindsay, M.W. Austin et al., A microwave channelizer and spectroscope based on an integrated optical Bragg-grating Fabry-Perot and integrated hybrid Fresnel lens system. IEEE Trans. Microwave Tech. **54**(2), 868-872P (2006)
9. D.B. Hunter, R.A. Minasian, Microwave optical filters using in-fiber Bragg grating arrays. IEEE Microwave & Guided Wave Lett. **6**(2), 103-105P (1996)
10. P.F. Snawerdt, M.D. Koontz, R.K. Morse et al., Acousto-optic channelizer-based ultra-wideband signal processor, U.S. Patent, 2000, 6091522
11. W. Wang, R.L. Davis, T.J. Jung et al., Characterization of a coherent optical RF channelizer based on a diffraction grating. IEEE Trans. Microwave Theory Tech. **49**(10), 1996–2001 (2001)

12. T. Sharpe, C. McCann, C. Harrington et al., Ultra-wideband cueing receiver based on spatial-spectral holographic rainbow spectrometer, in *Conference on Lasers and Electro-Optics* (IEEE, San Jose, 2014), pp. 1–2
13. X. Xie, Y. Dai et al., Broadband photonic radio-frequency channelization based on a 39-GHz optical frequency comb. IEEE Photonics Technol. Lett. **24**(8), 661–663 (2012)
14. A.O.J. Wiberg, D.J. Esman, L. Liu et al., Coherent filterless wideband microwave /millimeter-wave channelizer based on broadband parametric mixers. J. Lightwave Technol. **32**(20), 3609–3617 (2014)
15. K.O. Hill, G. Meltz, Fiber Bragg grating technology fundamentals and overview. J. Lightwave Technol. **15**(8), 1263–1276 (1997)
16. A. Yariv, Coupled-mode theory for guided-wave optics. IEEE J. Quantum Electron. **9**(9), 919–933 (1973)
17. D. Marcuse, *Theory of Dielectric Optical Waveguides* (Academic Press, New York, 1974)
18. X. Xie, Y. Dai, Y. Ji et al., Broadband photonic radio-frequency channelization based on a 39-GHz optical frequency comb. IEEE Photonics Technol. Lett. **24**(8), 661-663P (2012)
19. Y. Liu, L. Dong, J.J. Pan et al., Strong phase-controlled fiber Bragg gratings for dispersion compensation. Optics Lett. **28**(10), 786–788 (2003)
20. Y.W. Song, S.M.R. Motaghian, D. Starodubov et al., Tunable dispersion slope compensation for WDM systems using a single non-channelized third-order-chirped FBG, in *Optical Fiber Communication Conference and Exhibit (OFC)* (IEEE, Anaheim, 2002), pp. 580–581
21. G.C. Valley, Photonic analog-to-digital converters. Optical Soc. Am. **15**(5), 1955–1982 (2007)
22. G.C. Valley, J.P. Hurrell, G.A. Sefler, Photonic analog-to-digital converters: fundamental and practical limits. Proc. SPIE **5618**, 96–106 (2004)
23. H. Taylor, An optical analog-to-digital converter-design and analysis. IEEE J. Quantum Electron. **15**(4), 210–216 (1979)
24. R.G. Vaughan, N.L. Scott, D. Rod White, The theory of bandpass sampling. IEEE Trans. Signal Process. **39**(9), 1973–1984 (1991)
25. W.D. Gregg, *Analog and Digital Communications Systems* (Wiley, New York, 1977)
26. J. Delfyett Peter, C. DePriest, T. Yilmaz, Signal processing at the speed of lightwaves [photonic ADCs]. IEEE Circuits Devices Mag. **18**(5), 28–35 (2002)

Chapter 8
Optical Frequency Comb Generation Mechanism and Application

Optical frequency combs have been extensively studied by researchers and were the subject of work for which one half of the 2005 Nobel Prize in Physics was awarded. The performance of the optical frequency comb in the frequency domain is that the interval between each frequency component is equal, like the comb we use every day. Because of its unique frequency distribution, the optical frequency comb provides us with a bridge between optical frequency and radio frequency. The optical frequency comb converts unknown optical frequency information into radio frequency information and analyzes and controls unknown parameters by analyzing the radio frequency information. Its applications include optical frequency measurement, high-quality optical clock source generation, gas molecule measurement, ultra-fast optical signal processing and photon arbitrary waveform generation [1–4]. The use of optical frequency combs can organically integrate radio frequency and optical frequencies. In the field of metrology, the use of optical frequency combs can greatly improve measurement accuracy. In addition to traditional optical processing, optical frequencies are also widely used in optical fiber communication systems and other fields. In the field of optical communication, the use of optical frequency combs can provide channel references for bit-dense WDM communication networks. Depending on the application scenario, the method of generating an optical frequency comb is different.

8.1 Optical Frequency Comb Generation Method

The current methods of generating optical frequency combs are mainly based on femtosecond pulse mode-locked lasers to generate optical frequency combs, phase modulation-based Fabry–Perot cavity to generate optical frequency combs, optical fiber based on self-phase modulation in optical fibers and microresonance. The cavity generates an optical frequency comb and an optical frequency comb based on

J. Zhang and J. Li, *Satellite Photoelectric Sensing Technology*,
https://doi.org/10.1007/978-3-030-89843-4_8

intensity modulation and phase modulation. The following briefly introduces several generation techniques of optical frequency combs.

8.1.1 Generation of Optical Frequency Combs Based on Mode-Locked Lasers

In 1978, Hänsch and Chebotayev first realized the optical frequency comb using pulse trains with periodic repetition in the time domain [5]. Hänsch uses optical frequency combs from mode-locked lasers to measure Na. At the same time, Chebotayev also used the same method to measure the two-photon absorption resonance characteristics. Many types of mode-locked lasers have been developed so far, and one of them is now introduced—a saturable absorber mode-locked laser. The principle of the laser is shown in Fig. 8.1. A saturable absorber is a material that absorbs weak light and passes strong light. Therefore, when the light pulse passes through the saturable absorber, the part with high light intensity in the middle of the light pulse passes through the saturable absorber, and the part with weak light intensity at the edge of the light pulse is absorbed. The process was narrowed. By using saturable absorbers, high power narrow pulse width femtosecond pulse trains with passive mode locking can be realized. However, the repetition frequency of light pulses based on mode-locked lasers is generally tens to hundreds of MHz, that is, the free spectral range of the optical frequency comb is tens to hundreds of MHz, and its pulse coherence and pulse stability need to be improved.

When the frequency condition of the laser cavity meets the following formula

$$\omega_q - \omega_{q-1} = \frac{\pi c}{l} \equiv \Omega$$

the resonant cavity can achieve oscillation, where Ω is the mode interval of the oscillation, ω is the oscillation frequency, l is the cavity length, c is the speed of light and q is an integer. Then the complex form of the total light field of the multimode oscillation at any point in the end face of the optical cavity is

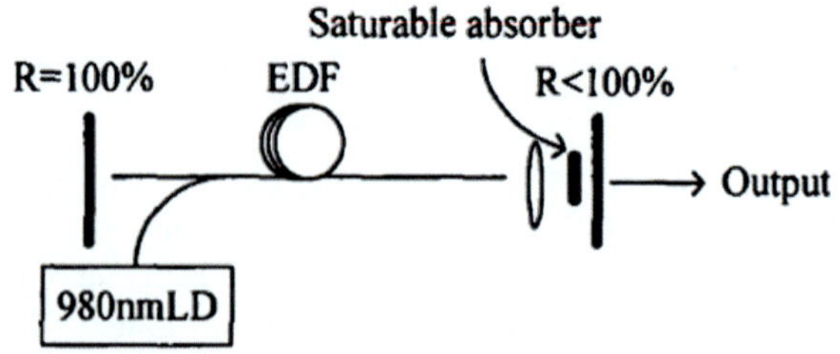

Fig. 8.1 Principle of saturable absorber mode-locked laser

$$E(t) = \sum_m C_m e^{i[(\omega_0 + m\Omega)t + \phi_m]}. \tag{8.1}$$

In the formula, C_m is the amplitude of the mth mode, ϕ_m is the phase of the mth mode, the summation number is the sum of the oscillation modes and ω_0 is an arbitrary reference frequency. It can be seen from Eq. (8.1) that $|E(t)|$ is a periodic function and its period is $\tau \equiv 2\pi/\Omega = 2l/c$, which is exactly equal to the time of one round trip in the resonant cavity. Using Eq. (8.1), we can get the field strength expression at time $t + \tau$:

$$\begin{aligned} E(t+\tau) &= \sum_m C_m \exp\left\{i\left[\left((\omega_0 + m\Omega)\left(t + \frac{2\pi}{\Omega}\right) + \phi_m\right)\right]\right\} \\ &= \sum_m C_m \exp\{i[((\omega_0 + m\Omega)t + \phi_m)]\} \exp\left\{i\left[2\pi\left(\frac{\omega_0}{\Omega}\right) + 2\pi m\right]\right\} \\ &= E(t)\exp(i2\pi\omega_0/\Omega). \end{aligned} \tag{8.2}$$

Since $E(t + \tau)$ is equal to $E(t)$, it can be seen from Eq. (6–2) that the two differ only by a constant phase factor. Therefore, we can analyze this result. If the fixed phase ϕ_m changes randomly over time, it will cause the laser output intensity to fluctuate randomly due to random interference between different modes. Therefore, in the early days of laser development, "mode-locking" technology has been proposed and demonstrated. Due to the mode-locking effect, the resonant cavity can generate a series of electromagnetic waves with $\tau = 2l/c = 2\pi/\Omega$ period, which means that this resonant cavity can generate an optical frequency comb. It can be seen that the basic principle of mode-locking is to lock the relative phase of each longitudinal mode that satisfies the vibration conditions in a certain form. The result from the time axis observation is that at some moments, all the oscillation modes reach the amplitude at the same time. At the largest position, the coherent superposition forms a periodic and extremely strong ultra-short pulse sequence, which appears in the frequency domain as an optical frequency comb with a pitch of the longitudinal mode.

Mode-locked lasers can be divided into active mode-locked, passive mode-locked, self-mode-locked and synchronous pump-locked modes according to the mode-locking mechanism.

8.1.2 Single Modulator Method

The method of generating an optical frequency comb based on a traditional Mach–Zehnder modulator (MZM) proposed by the Japanese Takahide Sakamoto research group provides a reference for the subsequent method of generating an optical frequency comb in terms of schemes and principles. The experimental schematic diagram is shown in Fig. 8.2.

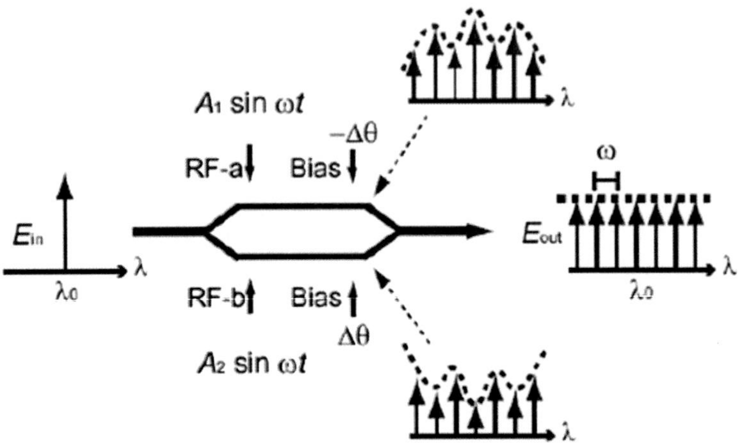

Fig. 8.2 Schematic diagram of optical frequency comb generation based on traditional MZM

Two RF signals are used to modulate the MZM. The two RF signals are

$$s_1(t) = A_1 \sin \omega t \tag{8.3}$$

$$s_2(t) = A_2 \sin \omega t. \tag{8.4}$$

When these two RF signals are modulated to MZM, the modulated optical signals can be expressed as

$$E_0 = \frac{E_{cw}}{2} \sum_k \left[J_k(B_1) e^{j(k\omega t + \theta_1)} + J_k(B_2) e^{j(k\omega t + \theta_2)} \right], \tag{8.5}$$

where E_{cw} is the light intensity of the input MZM, $B_1 = \pi A_1 / V_\pi$ and $B_2 = \pi A_2 / V_\pi$ are the modulation coefficients of the modulator, where A_1, A_2 are the amplitudes of the modulation signal, V_π is the half-wave voltage of the modulator, and θ_1 and θ_2 are the phase delays on each arm and $J_k(\bullet)$ is the k th expansion of the Bessel function. In the experiment, the added RF signal power is large. When the signal is large, the Bessel function can be approximated by the following formula:

$$J_k(B_i) \approx \sqrt{\frac{2}{\pi} B_i^{1/2}} \cos\left[B_i - \frac{(2k+1)\pi}{4} \right]. \tag{8.6}$$

Therefore, the conversion efficiency of each harmonic component of the output optical signal to the input optical signal can be equivalent to the following formula:

$$\eta_k = \frac{|E_k|^2}{|E_{cw}|^2}$$
$$= \frac{1}{2\pi \overline{B}}\left\{1 + \cos(2\Delta B) + [\cos(2\Delta\theta) + \cos(2\Delta\theta)]\cos\left(2\overline{B} - \frac{(2k+1)\pi}{2}\right)\right\}. \quad (8.7)$$

Among them, $\overline{B} = {}^{(B_1 + B_2)}/_2$, $\Delta B = {}^{(B_1 - B_2)}/_2$, $\Delta\theta = {}^{(\theta_1 - \theta_2)}/_2$ can be obtained from formula (8.7), when the following formula is satisfied:

$$\Delta B \pm \Delta\theta = {}^{\pi}/_2. \quad (8.8)$$

An optical frequency comb with flat comb teeth can be obtained. The formula (8.8) is called a flat comb condition.

When the RF signal power added to the MZM satisfies the flat condition of formula (8.8), for example, the two RF signal powers added are 35.9 and 36.4 dBm, respectively, and the phase difference $\Delta\theta$ is controlled by the variable delay line in the experiment. The final experimental results are shown in Fig. 8.3.

Figure 8.3(a) shows the result of only one arm plus RF signal modulation, and Fig. 8.3(b) shows the result of both arms plus RF signal modulation. It can be seen that this scheme can produce 11 optical frequency combs with flatness in the range of 1.1 dB.

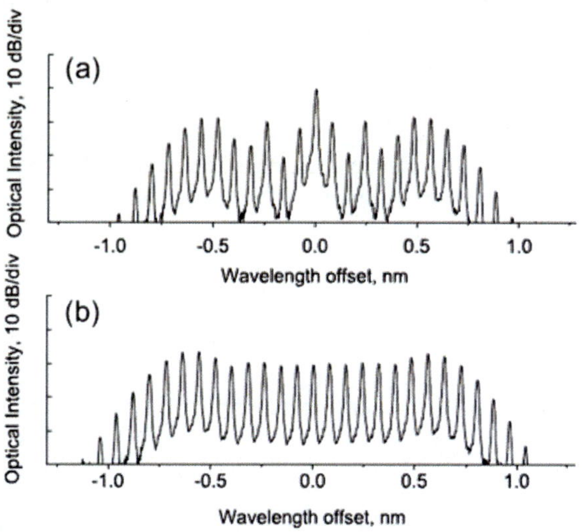

Fig. 8.3 Spectral plots of 11 flat optical frequency combs measured experimentally

8.1.3 Optical Frequency Comb Generation Based on Cascade of Intensity Modulation and Phase Modulation

In order to further simplify the generation of optical frequency combs, a method based on the cascade of intensity modulation and phase modulation has been proposed to generate optical frequency combs. This solution has the advantages of simple structure, stable operation, easy tuning and accurate frequency interval [6]. Figure 8.4 shows the generation of an optical frequency comb based on the cascade of intensity modulation and phase modulation. From the figure we can see that a continuous light is incident into the intensity modulator and the phase modulator. The two intensity modulators are responsible for flattening the optical frequency comb, while the phase modulator is mainly responsible for broadening the spectrum. PS is a microwave phase shifter, which is responsible for adjusting the microwave phase input to the three modulators. By adjusting the three microwave phase shifters, the flatness of the optical frequency comb can be adjusted. This solution can achieve a flat broadband, optical frequency comb with adjustable frequency interval, but this solution also has the disadvantage of a large volume of power consumption.

As shown in Fig. 8.5, PC1, PC2 and PC3 are polarization controllers, PolM is a polarization modulator, RF is a radio frequency signal generator, PS is a phase shifter, Pol is an analyzer, PM is a phase modulator and IM is an intensity modulator. Two sets of optical frequency combs with different frequency intervals are used as the multi-wavelength optical carrier source and the multi-wavelength optical local oscillator source. By designing the frequency interval of the two sets of optical frequency combs, the output of each channel has the same intermediate frequency. By sending the optical straight wave signal to be adjusted to the phase and intensity modulation module, the phase modulation coefficient is adjusted and the intensity is adjusted. The adjusted signal is sent to the polarization modulation module for polarization adjustment, and finally, the phase difference is corrected in the correction module. Polarization modulation angle correction outputs the optical signal of a flat optical frequency comb.

Send the optical straight wave signal to be adjusted to the phase and intensity modulation module and perform phase modulation and intensity modulation,

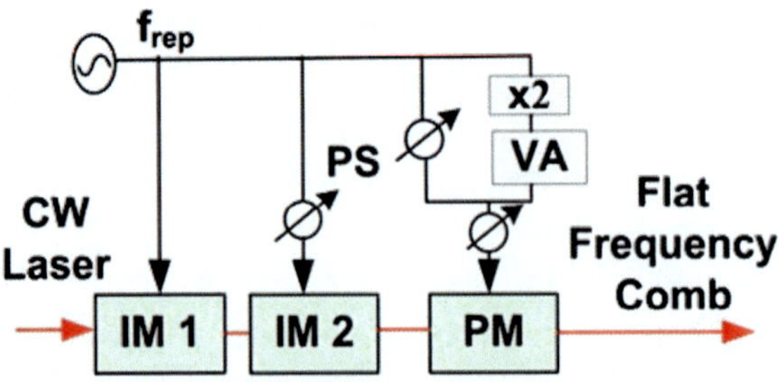

Fig. 8.4 Optical frequency comb generation scheme based on intensity modulation and phase modulation cascade

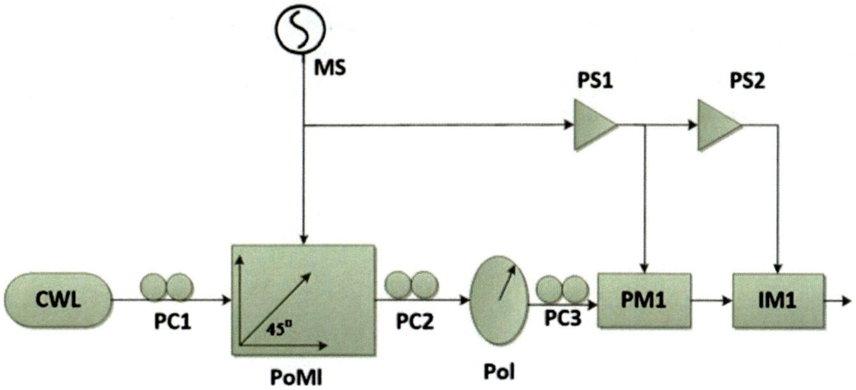

Fig. 8.5 Cascaded structure of polarization modulation module, phase and intensity modulation module

and send the modulated optical straight wave signal to the polarization modulation module.

The phase adjustment processing formula for the optical straight wave signal to be adjusted is

$$E_{out} = A \sum_{n=-\infty}^{\infty} J_n(\pi R) \exp[j E_{out}(\omega_0 + n\omega_1)t], \tag{8.9}$$

where E_{out} is the output signal of the sinusoidal signal loaded on the phase modulator and intensity modulator, A is the amplitude of the continuous wavelength laser output, ω_0 is the angular frequency of the continuous wavelength laser, ω_1 is the angular frequency of the RF-driven sinusoidal signal and $J_n(R)$ is the Bessel function.

The formula for adjusting the intensity of the straight wave signal to be adjusted is

$$\begin{aligned} E_{out} &= E_{in}(t) \exp(\pi(R_1 + R_2)\cos(\omega_1 t))\{1 + \exp[j\pi(\beta_{DC} + \beta_{IM}\cos\omega_1 t)]\} \\ &= E_{in}(t) \exp(\pi(R_1 + R_2)\cos(\omega_1 t)) \sin^2\left\{\frac{1}{2}(j\pi(\beta_{DC} + \beta_{IM}\cos\omega_1 t))\right\}, \end{aligned} \tag{8.10}$$

where E_{in} is the input signal, ω_1 is the angular frequency of the RF drive sinusoidal signal, β_{DC} and β_{IM} are the normalized values of the intensity modulator bias current and drive current amplitude to the DC half-wave voltage and AC half-wave voltage, and R_1 and R_2 are modulation coefficients of the phase modulator.

The polarization modulation module is used to perform polarization modulation on the received phase and intensity-modulated optical straight wave signal and sends the polarization-modulated optical straight wave signal to the correction module.

The specific steps of the polarization modulation module to perform modulation analysis on the optical signal spectrum output in step (1) are as follows:

(1) The formula for modulating the output optical signal spectrum by the polarization modulation module obtained according to the analysis results is as follows:

$$E_0 = \sqrt{P/2} \sum_{k=-\infty}^{\infty} J_k(\alpha_1) \exp(j(\omega_0 + k\omega_1)t)\left[\cos\theta_1 + (-1)^k \sin\theta_1 \exp(j\varphi_1)\right], \tag{8.11}$$

where P is the light intensity of the input signal, ω_0 is the angular frequency of the continuous-wavelength laser output, ω_1 is the angular frequency of the RF-driven sinusoidal signal, $\alpha_1 = \pi A / V_{\pi 1}$ is the degree of modulation, and θ_1 is the angle between the axis of the polarization modulator (PolM) and the main axis of the analyzer. $J_k()$ is the Bessel function.

(2) Calculate the light carrier intensity, ± first-order carrier intensity, and ± second-order carrier intensity of the output optical signal according to the modulation formula of the polarization modulation module. The expression is as follows:

$$E_{o0} = \exp(j\omega_0 t) J_0(\alpha_1)\left[\cos\theta_1 + \sin\theta_1 \exp(j\varphi_1)\right]$$

$$E_{o1} = -E_{o-1} = \exp[(j(\omega_0 \pm \omega_1)t)] J_1(\alpha_1)\left[\cos\theta_1 - \sin\theta_1 \exp(j\varphi_1)\right]$$

$$E_{o2} = -E_{o-2} = \exp[(j(\omega_0 \pm 2\omega_1)t)] J_2(\alpha_1)\left[\cos\theta_1 + \sin\theta_1 \exp(j\varphi_1)\right], \tag{8.12}$$

where ω_0 is the angular frequency of the continuous wavelength laser output, ω_1 is the angular frequency of the RF-driven sinusoidal signal, $\alpha_1 = \pi A / V_{\pi 1}$ is the modulation degree, and θ_1 is the angle between the polarization modulator (PolM) axis and the main axis of the analyzer, and $J_k()$ is the base Seoul function.

In the polarization modulation module, N polarization modulators need to be cascaded to finally output a flat optical frequency comb, where N is a positive integer. During the modulation process, the RF signal modulates the two vertical polarization states of the optical carrier in opposite phases, thereby introducing a phase difference φ. The optical signal passes the polarization controller through the analyzer output and also introduces one of the input polarized light. The angle θ between the axes is then output by the phase modulator and the intensity modulator to obtain a Bessel expansion model of the constructed spectral signal. There are six adjustable parameters $(\theta, \varphi, \alpha_1, R_{rf}, \beta_{DC}, \beta_{IM})$ in the final expression. Wavelength number output is more operable.

The phase difference correction is performed on the optical straight wave signal after the polarization modulation, and the polarization modulation angle correction is performed on the optical straight wave signal after the phase difference correction.

The corrected optical straight wave signal is output to achieve the flatness adjustment of the optical frequency comb.

The steps of correcting the phase difference and the polarization modulation angle of the optical straight wave signal after polarization modulation are as follows:

(3) The formula for correcting the phase difference of the optical straight wave signal after polarization modulation is

$$\begin{bmatrix} E_{x1}(t) \\ E_{x2}(t) \end{bmatrix} = \sqrt{2/2}\begin{bmatrix} \exp(j\omega_2 t) + j\alpha_1 \cos\omega_3 t \\ \exp(j\omega_2 t) - j\alpha_1 \cos\omega_3 t + j\varphi \end{bmatrix}, \tag{8.13}$$

where E_{x1} and E_{x2} are the light field strength components on two orthogonal axes, φ is the phase difference between the two orthogonal axes of the input polarized light, ω_0 is the angular frequency output by the continuous wavelength laser and ω_1 is the RF-driven sinusoidal signal angular frequency.

(4) The formula for correcting the polarization modulation angle of the optical straight wave signal after polarization modulation is

$$E_{o1} = \left[\exp(j\omega_2 t - j\alpha_1\omega_3 t)\sin\theta + \exp(j\omega_2 t + j\alpha_1\omega_3 t + j\varphi)\cos\theta\right] \tag{8.14}$$

$$E_{o2} = E_{o1}\exp\left(jR_{rf}\cos\omega_3 t\right)\sin^2\left\{\frac{1}{2}(j\pi(\beta_{DC} + \beta_{IM}\cos\omega_3 t))\right\}, \tag{8.15}$$

where θ is the angle between the main axis of the analyzer and one axis of the input polarized light.

The final output modulated by the phase modulator and intensity modulator is

$$E_{\text{out}} = \exp(j\omega_0 t)\sin^2\left\{\frac{1}{2}(j\pi(\beta_{DC} + \beta_{IM}\cos\omega_1 t))\right\}$$
$$\sum_{n=-\infty}^{\infty} j^n \exp(jn\omega_1 t)\left\{J_n\left(R_{rf} - \alpha_1\right)\sin\theta + J_n\left(R_{rf} + \alpha_1\right)\cos\theta\right\}, \tag{8.16}$$

where $J_n(\cdot)$ is the coefficient of the Bethel expansion, and there are six adjustable parameters $\left(\theta, \varphi, \alpha_1, R_{rf}, \beta_{DC}, \beta_{IM}\right)$ in the final expression.

The following further describes in combination with specific embodiments:

θ	φ	α_1	R_{rf}	β_{DC}	β_{IM}	Flatness(dB)
5.47	0.58	3.60	1.76	0.72	0.35	0.5
3.89	2.49	3.64	1.87	0.69	0.54	1.37
0.64	1.08	4.93	1.82	0.57	0.64	2.29
0.88	0.59	1.98	5.49	0.74	0.66	2.78
3.88	2.52	0.42	8.0	0.79	0.63	1.80

(continued)

(continued)

θ	φ	α_1	R_{rf}	β_{DC}	β_{IM}	Flatness(dB)
5.53	5.69	0.39	9.14	0.85	0.33	1.65

8.1.4 Generation of Optical Frequency Combs Based on Phase-Modulated Optical Resonator

Immediately after the mode-locked laser, people began to use phase-modulated Fabry–Perot cavities to generate optical frequency combs. Kourogi of the Tokyo Institute of Technology in Japan placed a 6 GHz electro-optic modulator in an optical cavity so that the light incident on the optical cavity was repeatedly frequency-shifted, resulting in an optical frequency comb with a frequency interval of 6 GHz, and its frequency coverage up to a few terahertz [7]. The free spectral range of the optical frequency comb based on the phase-modulated optical resonator is determined by the frequency of the loaded microwave signal, so the accuracy of the spacing between adjacent sidebands is determined by the frequency accuracy of the microwave source. Therefore, the frequency accuracy of this optical frequency comb is very high. Kourogi uses a bait-doped fiber amplifier to amplify the optical pulse, thereby generating an optical frequency comb with an average power of large watts, which provides the possibility to further expand the frequency coverage of the optical frequency comb to 50 THz by using four-wave mixing. The optical frequency generation scheme proposed by Kourogi is used to measure the step frequency of hydrogen energy and to measure the accuracy of the optical frequency combs generated by femtosecond mode-locked lasers. Although the frequency coverage of the optical frequency comb is now as high as THz, compared to the optical carrier frequency, the frequency coverage is only 5% of the optical carrier frequency. Therefore, people need to further broaden the frequency comb, but the optical frequency comb generation mechanism based on the optical resonator is difficult to meet the requirements because the specular reflection of the optical resonator has a bandwidth limitation, so it is difficult to achieve a wider frequency coverage.

8.1.5 Generation of Optical Frequency Combs Based on Self-Phase Modulation in Optical Fibers

In order to overcome the above difficulties, Imai proposed an optical frequency comb generation scheme based on fiber self-phase modulation. As shown in Figure 8.6 [8], a fiber amplifier is used to amplify the generated optical frequency comb, thereby exciting the self-phase modulation of the optical frequency comb in the fiber. Since

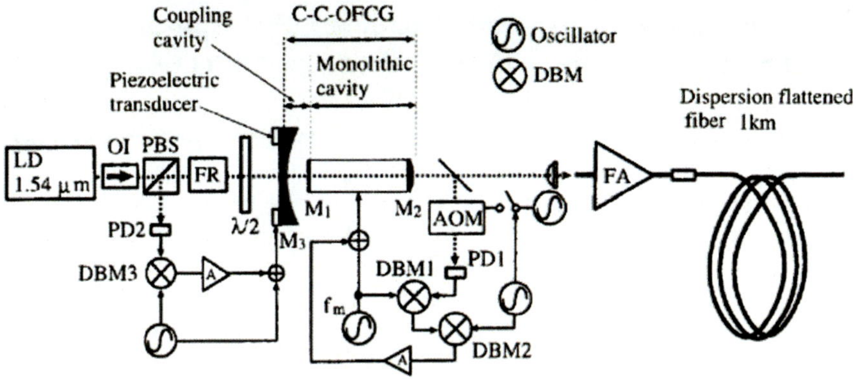

Fig. 8.6 Optical frequency generation scheme based on fiber self-phase modulation

self-phase modulation is a third-order nonlinear process, more frequency components can be generated by the four-wave mixing effect. The frequency components generated by four-wave mixing are the "sum" of other frequency components. As long as the conservation of energy is satisfied, the frequency components of the momentum conservation and dispersion equations can produce a four-wave mixing effect. Therefore, the frequency coverage of the optical frequency comb can be effectively extended by four-wave mixing. However, this scheme needs to meet the four-wave mixing conditions strictly, and the whole scheme is difficult to debug.

8.1.6 Generation of Optical Frequency Combs Based on Micro-Resonant Cavity

In order to achieve a more compact and integrated generation of optical frequency combs, Del'Haye P. proposed an optical frequency comb generation scheme based on Kerr nonlinearity and a microresonator. This solution can generate an optical frequency comb with frequency coverage of up to 70 THz without resorting to an additional frequency stretching mechanism. Compared with other schemes, the use of a micro-resonator cavity provides the possibility for an integrated and miniaturized optical frequency comb generation, which fundamentally reduces the volume power consumption and price of the optical frequency comb generation scheme. This scheme is also a solution for future optical frequency combs [9].

8.2 Experiments to Generate Broadband Flat Optical Frequency Combs with High-Frequency Intervals

The different schemes for generating optical frequency combs are described above. According to different application scenarios, the schemes for generating optical frequencies are also different. For example, for optical frequency measurement, the frequency coverage of the required optical frequency comb needs to be one octave higher; for optical detection, the frequency interval of the required optical frequency comb is tens of MHz; and for optical fiber communication, in the field of microwave photonics, the frequency interval of the required optical frequency comb needs to be as high as several tens of GHz. In order to generate broadband flat optical frequency combs with high-frequency intervals, we have studied and experimented on the generation mechanism of broadband flat optical frequency combs based on recirculating frequency shifter (RFS). Below the optical frequency comb generation method is introduced in detail.

8.2.1 Generation of Broadband Flat Optical Frequency Combs Based on RFS

Optical frequency combs are now widely used in arbitrary waveform generation, microwave photon filters, dense wavelength division multiplexing and orthogonal frequency-division multiplexing systems. Many methods have been proposed to generate optical frequency combs, but the flatness and bandwidth of optical frequency combs need to be improved. We achieve a broadband flat optical frequency comb by means of cyclic frequency shift. The frequency interval of the generated optical frequency comb is 20 GHz, and the frequency coverage is >1.6 THz. The flatness of the entire optical frequency comb reaches 2.5 dB.

8.2.2 Principle

Figure 8.7 is a schematic diagram of an RFS-based optical frequency comb scheme. We can see that the whole solution is composed of laser, optical coupler, polarization controller, bait-doped fiber amplifier, I/Q modulator, microwave phase shifter and microwave amplifier. CW is a continuous light source, PC is a polarization controller, PS microwave phase shifter, IQM is an I/Q modulator and Erbium-doped fiber amplifier (EDFA) is a bait-doped fiber amplifier. The principle of this solution is as follows: The continuous light generated by the laser is divided into two channels by an optical coupler, one of which is used as an output end, and the other is input to an I/Q modulator. By adjusting the bias of the I/Q modulator to make it in the single-sideband carrier suppression state, after passing through the I/Q modulator,

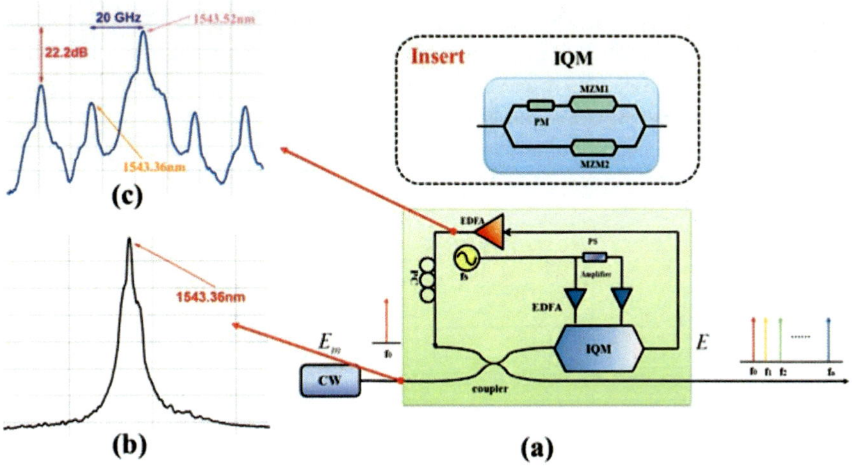

Fig. 8.7 Generation of RFS-based broadband flat optical frequency comb

the optical frequency comb is moved unidirectionally. Then by repeating the unidirectional frequency shift, an optical frequency comb is generated. The bait-doped fiber amplifier is used to compensate for the loss of the entire loop. The polarization controller aligns the polarization state of the entire loop. The two microwave power amplifiers are in the saturated output state, and the output RF power is 26 dBm, so the power input to the two RF ports of the I/Q modulator is the same. The microwave phase shifter is used to adjust the phase difference of the microwave signals reaching the two RF ports of the I/Q modulator. In the experiment, we need to ensure that the phase difference of the microwave signals of the two input ports is 90°.

The incident continuous optical signal can be expressed by the following expression:

$$E_{in} = E_0 \exp(j2\pi f_0 t), \tag{8.17}$$

where f_0 is the frequency of the incident light and E_0 is the amplitude of the incident light. The spectrum of the input continuous light is shown in Fig. 6.7(b). The incident light is divided into two ways by the coupler, and one of them enters the I/Q modulator. The I/Q modulator is cascaded with two Mach–Zehnder modulators and a phase modulator in parallel. Its structure is shown in the illustration in Fig. 8.7. By using a microwave phase shifter, the signals at the two RF input terminals of the I/Q modulator are out of phase by 90°. The two signals can be expressed as

$$V_I = V_{DCI} + V_{PPI} \cos(2\pi f_s t) \tag{8.18}$$

$$V_Q = V_{DCQ} + V_{PPQ} \sin(2\pi f_s t). \tag{8.19}$$

V_{DCI} and V_{DCQ} are the DC bias voltages of the two RF ports, V_{PPI} and V_{PPQ} are the peak-to-peak voltages of the microwave signals of the two RF ports and fs is the frequency of the microwave signal and the frequency interval of the optical frequency comb. By adjusting the bias voltage on the I/Q modulator to make it in the carrier-suppressed single-sideband modulation mode, the output light field of the I/Q modulator can be expressed as

$$\begin{aligned} E_{out} &= \frac{E_0}{2}\exp(j2\pi f_0 t)\left[j\sin\left(\frac{\pi}{2}\frac{V_{PP}\sin(2\pi f_s t)}{V_\pi}\right) + \sin\left(\frac{\pi}{2}\frac{V_{PP}\cos(2\pi f_s t)}{V_\pi}\right)\right] \\ &\approx J_1\left(\frac{\pi V_{PP}}{2V_\pi}\right)E_0\exp[j2\pi(f_0 + f_s)t]. \end{aligned} \tag{8.20}$$

The spectrum of the optical signal obtained after passing through the I/Q modulator is shown in Fig. 6.7(c). From Eq. 6.20, we can see that after passing through the I/Q modulator, the frequency of the incident light shifts f_s to high frequencies. Then after one cycle, the output light in the entire RFS can be expressed as

$$E_1 = E_{in} + E_{out} = E_0\exp(j2\pi f_0 t) + J_1\left(\frac{\pi V_{PP}}{2V_\pi}\right)E_0\exp[j2\pi(f_0 + f_s)t]. \tag{8.21}$$

From Eq. 8.21, we can see that after one cycle, the RFS output light contains two frequency components, f_0 and $f_1 = f_0 + f_s$. Similarly, after two cycles, the output light will contain the frequency component of $f_2 = f_0 + 2\,fs$. Therefore, after N cycles, the frequency component of the output light of the RFS will include $f_0, f_1, f_2, \ldots, f_0 + Nf_s$. At this time, the output light is a broadband flat optical frequency comb with a high-frequency interval. The optical frequency comb produced in this way has an adjustable free spectral rånge and an ultra-wide flat optical frequency comb.

8.2.3 Experimental Results and Discussion

In the experiment, we used a narrow line width low-noise laser as the light source. The spectrum of the light source is shown in Fig. 8.7(b). Its wavelength is 1543.36 nm and the power is 8.84 dBm. The light source is divided into two by a 2 × 2 optical coupler, one of which is input to the I/Q modulator, and the other is used as an output port. The optical carrier wave input to the I/Q modulator is loaded with a microwave signal with a frequency of 20 GHz. We adjust the bias of the I/Q modulator to make the modulator in a carrier-suppressed single-sideband modulation state. Theoretically, the bias voltages of I and Q channels should be adjusted to the lowest bias points of the transmission curves of the two Mach–Zehnder modulators. At the same time, the bias

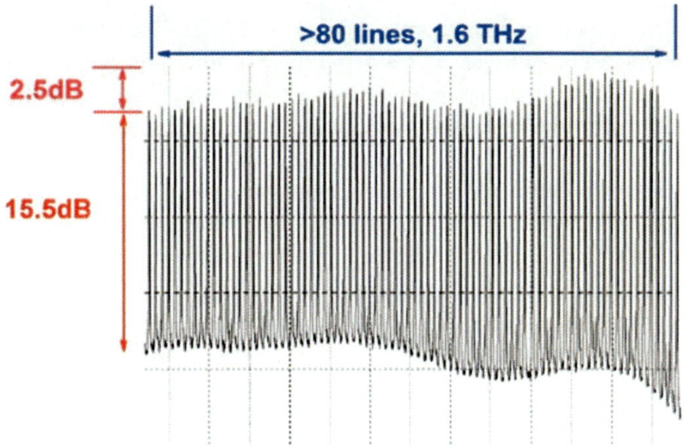

Fig. 8.8 Spectral plot of the optical frequency comb generated by the experiment

voltage of the phase modulator inside the I/Q modulator should be adjusted to 90°. The optical carrier phase of the upper and lower arms of the I/Q modulator is 90° out of phase, thereby realizing a single-sideband carrier suppression modulation method. In order to keep the power of the two microwave input ports of the I/Q modulator as consistent as possible, we use two power amplifiers with the same saturation power to saturate the input microwave signal. In the single-sideband carrier suppression modulation mode, after passing through the I/Q modulator for the first time, the input optical carrier shifted to 20 GHz in the long-wavelength direction, and the sideband suppression ratio reached 22.2 dB. The spectrum at the output of the I/Q modulator is shown in Fig. 8.7 (c). The loss through the modulator is 11 dB, so we use EDFA to compensate for the insertion loss.

After passing through the I/Q modulator N times, a large number of optical frequency components with an interval of 20 GHz can be generated. The spectrum is shown in Figure 8.8. The frequency coverage of the optical frequency comb generated by experiments exceeds 1.6 THz, the unevenness of the entire frequency comb is 2.5 dB and the signal-to-noise ratio of the entire optical frequency comb is >15.5 dB.

8.3 Generation Technology of Bi-Coherent Optical Frequency Comb Based on Time Lens Method

A key issue for channelization using signal multicast is the generation of a multi-wavelength light source that acts as a stable frequency interval between the carrier and the local oscillator, has a large number of wavelengths and has a high degree of flatness. At present, there are many ways to achieve this. The super-continuous spectrum vertical mode segmentation technology uses high-linear fiber to broaden the pulsed light spectrum and then uses a comb filter to filter and shape the frequency spectrum. The multi-mode laser light source technology is to set a filter with a limited band inside the resonant cavity as required. Unit, selectively controlling the output wavelength; and optical frequency comb technology using micro-ring resonator or electro-optic modulation. Among them, since the comb spectrum was researched and produced by T.W. Hnnsch and others at Stanford University in 1977, the optical frequency comb has become a research hotspot as a new light source, which has attracted widespread attention. Relying on its multi-carrier structure, the frequency interval of the tooth line is fixed, which can achieve more than tens of GHz, the coherence between the comb teeth is good and any two tooth lines can be beaten to generate radio frequency signals. Advantages include precision instrument calibration and astronomy. Observation, ranging and signal processing are rapidly becoming widespread. There are many generation schemes of optical frequency combs. The mode-locked laser generation technology is based on gain fiber, the Kerr optical frequency comb generation technology is based on microcavity, the cascaded modulator method is based on time lens and the optical frequency comb is based on cyclic frequency shifter. The generation technology of a large number of tooth lines and the generation technology based on nonlinear effects such as self-phase modulation and four-wave mixing etc., in actual applications, different generation mechanisms can be selected according to specific needs. Among them, the time lens-based generation scheme has become one of the most mature technologies due to the advantages of flexible and adjustable comb interval and a simple structure. This section will introduce the theoretical mechanism of this scheme in detail and set up an experimental device for dual-coherent optical frequency combs and test its performance.

8.3.1 Principle of the Optical Frequency Comb Generated by the Time Lens Method

The concept of the time lens was first proposed by BH Kolner and M. Nazarathy in 1989, and Kolner found the perfect duality of space–time from the Maxwell equations [128], that is, the propagation characteristics of the pulse in time in a dispersion medium are similar to those in space. The diffraction characteristics of the beam and the secondary phase modulation with respect to time in the time domain are similar to the secondary phase modulation with respect to distance in the space domain. Under paraxial diffraction, the light field propagation function in free space is

$$\frac{\partial E}{\partial z} = -\frac{i}{2k}\left(\frac{\partial E^2}{\partial x^2} + \frac{\partial E^2}{\partial y^2}\right), \tag{8.22}$$

where E is the intensity of the light field, k is the propagation constant and z is the direction of propagation. In the time domain the pulse is transmitted in a dispersive medium

$$\frac{\partial A}{\partial z} = \frac{i\beta_2}{2} \cdot \frac{\partial A^2}{\partial t^2}, \tag{8.23}$$

where A is the pulse amplitude, β_2 is the group velocity dispersion constant and t is time. The additional secondary phase of the light field in the frequency domain is

$$\phi(k_x) = -\frac{k_x^2 z}{2k}, \tag{8.24}$$

where k_x is the transverse wave number of the light wave. Similarly, the additional secondary phase of the pulse in the frequency domain can be expressed as

$$\phi(\omega) = -\frac{\beta_2 z \omega^2}{2}. \tag{8.25}$$

By comparing (8.22) and (8.23), (8.24) and (8.25), it can be found that both space and time are similar in both the time domain and the frequency domain, as shown in Figure 8.9. This similarity is called spatio-temporal correspondence. Analogous to a space lens, a device that introduces secondary phase modulation in time is called a time lens. The time lens can not only perform time-domain compression and stretch imaging of pulses but also realize real-time Fourier transform.

The most common component of a time lens is an electro-optic phase modulator. According to the time–frequency mapping, the top flatness of the time domain signal corresponds to the flatness of the tooth line of the optical frequency comb. In order to increase the number of tooth lines, two cascaded phase modulators are used here to generate a bi-coherent optical frequency comb. Assume that the optical carrier signal is

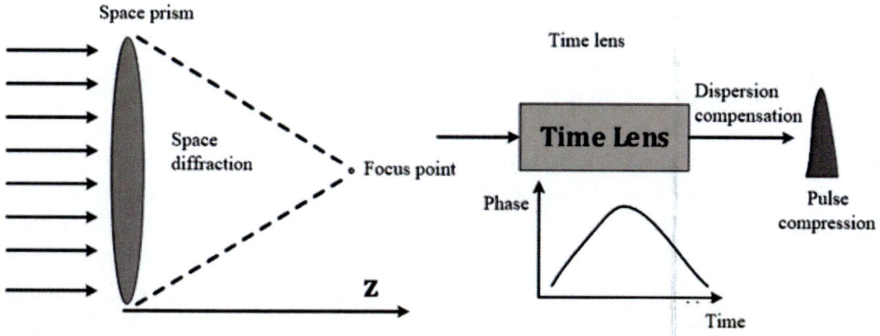

Fig. 8.9 Schematic diagram of time–space correspondence

$$E_{in}(t) = \sqrt{P_0} \exp j(\omega_0 t + \phi(t)), \tag{8.26}$$

where P_0 is the input optical power and $\phi(t)$ is the phase noise of the laser. Select the sine wave $v(t)$ as the modulation signal:

$$v(t) = V_0 \sin[\omega_1 t + \varphi_1(t)], \tag{8.27}$$

where V_0 is the amplitude of the modulation signal and $\varphi_1(t)$ is the phase noise of the sine wave. The light field of the output signal of the time lens is

$$E_{out}(t) = \sqrt{P_0} \exp j\{\omega_0 t + 2\beta \sin[\omega_1 t + \varphi_1(t) + \phi(t)\}, \tag{8.28}$$

where $\beta = V_0/V_\pi$ is the modulation index of the modulator and V_π is the half-wave voltage of the modulator. Perform the first kind of nth-order Bessel function expansion on the above formula:

$$E_{out}(t) = \sqrt{P_0} \sum_{n=-\infty}^{\infty} J_n(2\beta) \exp j\{\omega_0 t + n\omega_1 t + \phi(t) + n\varphi_1(t)\}. \tag{8.29}$$

It can be seen from the above formula that the output spectrum of the time lens is centered on ω_0, extended to both sides at intervals of ω_1, and the output is an equally spaced comb-shaped tooth line, that is, an optical frequency comb. The frequencies of the two coherent optical frequency combs can be respectively expressed as

$$f_{sig}(k) = f_{sig}(1) + (k-1)\delta_{sig} \tag{8.30}$$

$$f_{lo}(n) = f_{lo}(1) + (n-1)\delta_{lo}, \tag{8.31}$$

where $f_{sig}(1) = \omega_0/2\pi$ and $f_{lo}(1) = f_{sig}(1) + F_0$ are the central tooth line frequencies of the optical comb, $\delta_{sig} = \omega_1/2\pi$ and $\delta_{lo} = \omega_2/2\pi$ are the free spectral ranges, and n and k are the tooth line sequences. It can be known from (8.31) that the output signal is closely related to the modulation depth of the modulator, and its influence is mainly reflected in the power jitter of the output frequency of the optical frequency comb, that is, flatness. We simulated the optical frequency combs of the cascaded two-phase modulators. Taking the 10 GHz tooth line interval as an example, the simulation results are shown in Fig. 8.10. Figure 8.10(a) shows the number of optical tooth comb output tooth lines and the modulator. The relationship between the modulation depths shows that as the modulation depth increases, the number of tooth lines in the 5 dB range is increasing, and the flatness of the optical frequency comb is getting better and better. At present, the half-wave voltage level of the phase modulator cannot exceed 3 V. With a high-power RF drive of 1 W, the modulation

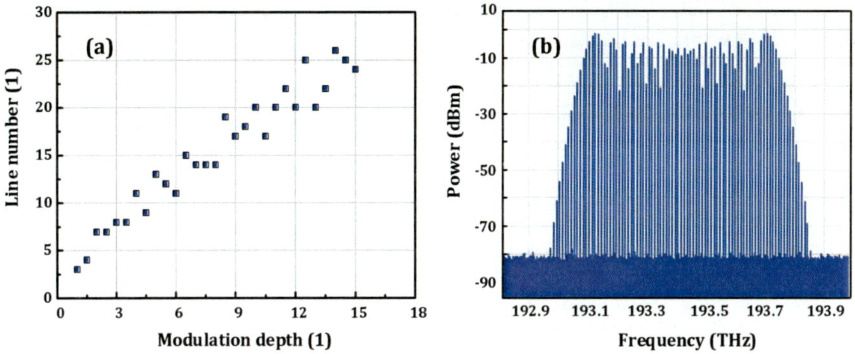

Fig. 8.10 **a** The relationship between the number of tooth lines and the modulation depth in the flatness range of 5 dB. **b** Simulation results of the optical frequency comb generated by the cascaded phase modulator at a modulation depth of 6

depth can reach about 6, and the number of tooth lines in the 5 dB flatness range can be more than ten. Cascade modulation of the number of frequency comb tooth lines generated by the device can reach more than 20. Figure 8.10(b) is the spectrum diagram of the optical frequency comb under this simulation condition.

8.3.2 Experimental Device for Generating Coherent Optical Frequency Combs

As shown in Fig. 8.11, the continuous light is first divided into two upper and lower paths by an optical coupler, and the upper path enters two cascaded phase modulators. The RF signal RF1 is divided into two parts by a power divider. For the modulator, a phase shifter is used for phase matching between the two RF signals to ensure that the phase difference between the two RF signals input to the modulator is zero. The final modulated optical signal is an optical frequency comb with a tooth line interval of RF1. The lower light first passes through a modulator driven by the frequency F0 and then passes through an optical filter to filter out a sideband as the seed light for generating a second coherent optical frequency comb. This step is to adjust the initial frequency difference from the upper light, thereby increasing the interval between the tooth lines of the two optical frequency combs. The seed light is amplified by EDFA and then passed through two modulators to generate a second optical frequency comb. The free spectral range of the two optical combs can be changed by adjusting RF1 and RF2.

Figure 8.12(a) shows the experimental setup built in the experiment. In addition, a comb-shaped optical filter (one input and two output devices) can be used to separate the optical frequency comb odd and even tooth lines to achieve double the free spectral range. Figure 8.12(b) is measured using a spontaneous emission noise source. Filter

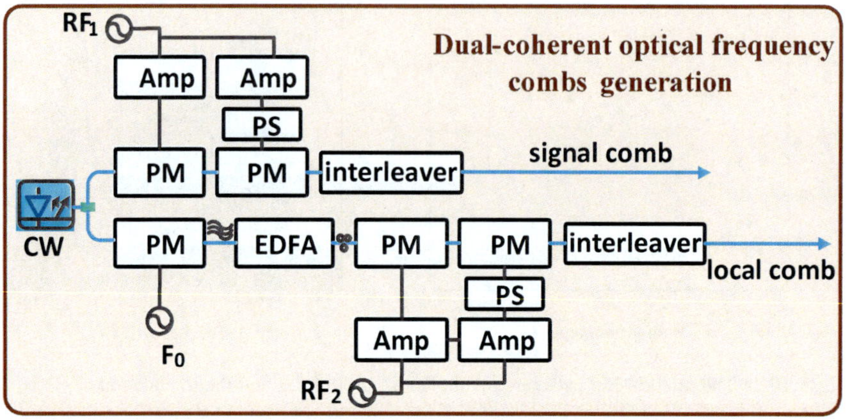

Fig. 8.11 The implementation structure diagram of generating a bi-coherent optical frequency comb. CW: continuous light laser; PM: phase modulator; RF1, RF2, F0: microwave source; Amp: radio frequency power amplifier; PS: radio frequency phase shifter

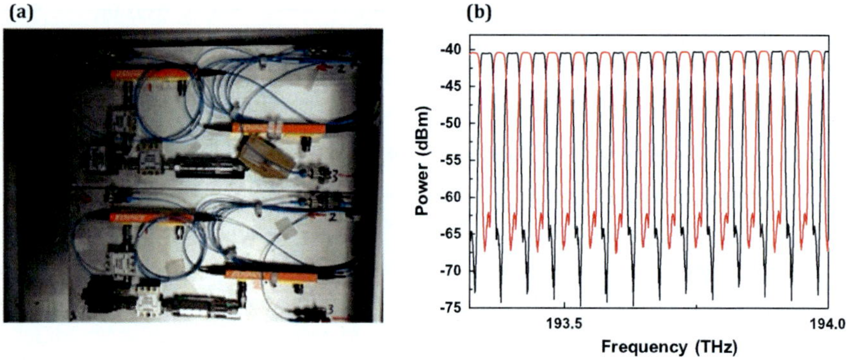

Fig. 8.12 **a** Experimental setup of a dual-coherent optical frequency comb. **b** Frequency response characteristics of comb filters

shape, the red and black curves are the frequency response characteristics of the odd and even channels, respectively.

8.3.3 *Experimental Results and Discussion*

In the experiment, the modulation frequencies of the two modulators are set to RF1 = 24.5 GHz and RF2 = 25.5 GHz, the power is 1 W through the RF amplifier, and the input optical power is 12 dBm. Figure 8.13(a) is the output of the signal optical frequency comb spectrum, the number of tooth lines in the 5 dB range is 23 and the

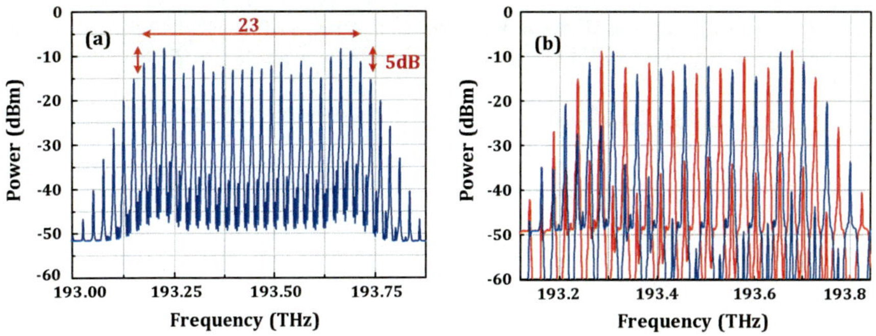

Fig. 8.13 Optical frequency comb produces experimental results. **a** Output optical frequency comb spectrum. **b** Output spectrum of filter odd and even channel

flatness is good. Figure 8.13(b) shows the two outputs of the optical frequency comb after passing through the comb filter. The red curve is the frequency spectrum of the odd-numbered tooth line, and the blue line is the frequency spectrum of the even-numbered tooth line. The flatness of the optical frequency comb is not only related to the modulation depth but also to the time domain shape of the modulation signal and the phase difference between the two drive signals of the cascaded modulator. The higher the flatness of the modulation signal, the greater the flatness of the frequency comb, the larger the phase difference, and the more uneven the frequency comb. Moreover, the center frequency appears asymmetric. When the phase difference is 0, the optical comb is the most flat, so usually the comb performance is best when using cascade phase modulation and intensity modulator together. In summary, although the time lens-based solution is relatively expensive, it has the most features such as simple operation, flexible and controllable parameters, and stable performance.

References

1. S.A. Diddams, D.J. Jones, J. Ye et al., Direct link between microwave and optical frequencies with a 300THz femtosecond laser comb. Phys. Rev. Lett. **84**(22), 5102-5105P (2000)
2. S.A. Diddams, T. Udem, J.C. Bergquist et al., An optical clock based on a single trapped $^{199}hg^{+}$ Ion. Science **293**(5531), 825-828P (2001)
3. M.J. Thorpe, K.D. Moll, R.J. Jones et al., Broadband cavity ringdown spectroscopy for sensitive and rapid molecular detection. Science **311**(5767), 1595-1599P (2006)
4. S.A. Diddams, L. Hollberg, V. Mbele, Molecular fingerprinting with the resolved modes of a femtosecond laser frequency comb. Nature **445**(7128), 627-630P (2007)
5. Y.V. Baklanov, V.P. Chebotayev, Narrow resonances of two-photon absorption of super-narrow pulses in a gas. Appl. Phys. **12**(1), 97-99P (1977)
6. R. Wu, V.R. Supradeepa, C.M. Long et al., Generation of very flat optical frequency combs from continuous-wave lasers using cascaded intensity and phase modulators driven by tailored radio frequency waveforms. Opt. Lett. **35**(19), 3234-3236P (2010)

7. M. Kourogi, K. Nakagawa, M. Ohtsu, Wide-span optical frequency comb generator for accurate optical frequency difference measurement. IEEE J. Quantum Electron. **29**(10), 2693-2701P (1993)
8. K. Imai, M. Kourogi, M. Ohtsu, 30 THz span optical frequency comb generation by self-phase modulation in an optical fiber. IEEE J. Quantum Electron. **34**(1), 54-60P (1993)
9. Del'Haye P., Schliesser A., Arcizet O., et al, Optical frequency comb generation from a monolithic microresonator, Nature, 2007, 450: 1214–1217P.

Chapter 9
Channelized Receiving Technology Based on Optical Frequency Comb

9.1 Channelized Filtering Receiving Technology Based on Fabry–Perot Filter

Future applications require military RF systems that can handle higher frequencies and greater bandwidth. However, due to the volume and power consumption of traditional RF devices, real-time, high-precision radio frequency spectrum detection cannot be achieved. Therefore, it is very important to divide the broadband RF signal into sub-channels compatible with modern digital systems. Because photon technology has the advantage of broadband, it has been used to process broadband RF signals.

Recently, some innovative photon schemes have been proposed to channelize RF signals. Most schemes are implemented through the structure of up-conversion channel division. This structure first modulates radio frequency signals onto optical carriers and is then divided into N channels by N physically separate filters. This spectral division can be achieved by Bragg gratings [1, 2], acousto-optic crystals [3] and integrated optical devices [4, 5]. But it should be noted that the bandwidth of each sub-channel is limited by the electronics. Therefore, a large number of narrow band, dense spectrum and stable optical filters are required. Reference [6] reports a solution to avoid this complicated filter design. The RF signal is copied to a non-coherent source of spectrum division. The spectrally divided incoherent source is generated by a broad-spectrum light source through a Fabry–Perot filter. The free spectral range of the second Fabry–Perot filter is slightly different from the free spectral range of the first Fabry–Perot filter. Each sub-channel is then divided into different channels by an optical demultiplexer. However, it is difficult for this scheme to perform coherent reception due to large noise and loss. Other photon channelization schemes are implemented by modulating the RF signal on multiple free-running lasers, or by parametric multicasting the RF signal through a highly nonlinear fiber [7]. However, these two solutions require precise frequency alignment, which increases the difficulty of practical operation. In order to overcome the above disadvantages, we propose that the channelization of the above-mentioned photon radio frequency signal can be achieved

J. Zhang and J. Li, *Satellite Photoelectric Sensing Technology*,
https://doi.org/10.1007/978-3-030-89843-4_9

by a broadband optical frequency comb. The frequency comb has high flatness and a high signal-to-noise ratio, which provides the possibility for low-noise channelization, simplifying the frequency alignment of subsequent filters. Below we discuss the frequency response and channelization characteristics of the scheme in theory and experiments.

9.1.1 Fabry–Perot Filter Principle

Now Fabry–Perot filter (FPF: Fabry–Perot filter) has been widely used in various fields of scientific research, such as precision test instruments, sensing applications, dense wavelength division multiplexed optical fiber communication systems, etc. Now there are types of Fabry–Perot filters such as metal-dielectric thin-film light filters, all-dielectric filters, multi-cavity thin-film filters and all-fiber. The Fabry–Perot filter was invented in 1879, after which Charles Fabry developed the multi-beam interference theory of light to theoretically analyze the principle of the Fabry–Perot filter. The biggest advantage of FPF is that it has a narrow line width that other optical filters cannot match. The filter bandwidth of modern general optical filters is ~GHz, while the filter bandwidth of FPF can be as narrow as 100 MHz. In addition to this, the periodic filtering characteristic of FPF is also widely used in dense wavelength division systems. In our experiments, we fully utilize the two characteristics of FPF narrow line width and periodic filtering to achieve channelization filtering. Below we briefly introduce FPF through the basic theory of multi-beam interference theory, as shown in Fig. 9.1.

The structure of Fabry–Perot filter is shown in Fig. 9.1. When the incident light enters the filter, it will be reflected multiple times between two parallel high reflectivity mirrors, and there is a fixed phase difference between adjacent reflected light.

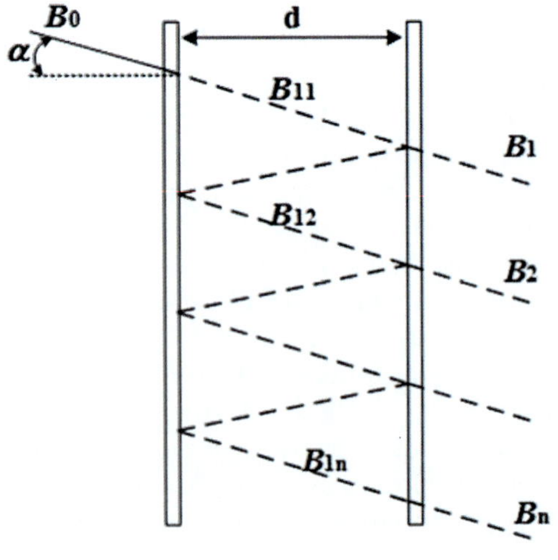

Fig. 9.1 Schematic diagram of the Fabry–Perot filter

Therefore, the light obtained by the Fabry–Perot filter is obtained. Periodic modulation forms a periodic transmission spectrum. Assume that the incident angle of the light beam to the FPF is a, the thickness of the two parallel mirrors is d, the reflectivity of the mirrors is R, the amplitude of the incident light wave is B0 and the incident light is reflected by the second piece of the Fabry–Perot filter. After the mirror, its amplitude is

$$\begin{aligned} B_1 &= \sqrt{1-R} \cdot R \cdot B_0 \\ B_2 &= \sqrt{1-R} \cdot R^2 \cdot B_0 \\ &\cdots \\ B_n &= \sqrt{1-R} \cdot R^n \cdot B_0 \end{aligned} . \tag{9.1}$$

The fixed phase difference of the light beam after passing the second mirror is

$$\delta = \frac{2kd}{\cos a}, \tag{9.2}$$

where $k = 2\pi/\lambda$, and k is the light wave vector. The expression of the final total emitted light intensity through interference between different beams is

$$I = I_0 \frac{(1-R)^2}{(1-R)^2 + 4 \cdot \sin^2(\delta/2)}. \tag{9.3}$$

By analyzing the above formula, we can analyze and obtain the free spectral range of FPF as

$$FSR = \lambda^2/(2nd). \tag{9.4}$$

The 3 dB bandwidth of FPF is

$$BW = \frac{c}{2\pi d} \cdot \frac{1-R}{\sqrt{R}}. \tag{9.5}$$

In our practical application, an all-fiber Fabry–Perot filter is used. Its free spectral range is about 40 GHz, which can be fine-tuned through temperature control. The 3 dB bandwidth is 400 MHz.

9.1.2 *Channelization Filtering Principle*

Our proposed channelization scheme based on optical frequency comb is shown in Fig. 9.2a. An optical frequency comb with a channel spacing of δ_{OFC} is modulated by a Mach–Zehnder modulator on a wideband RF signal, and the RF optical carrier is

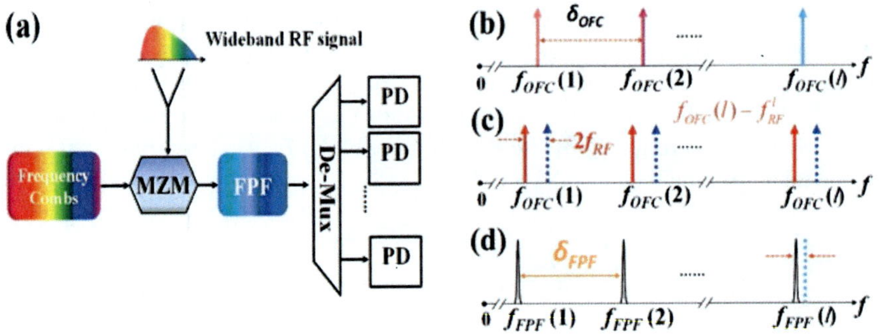

Fig. 9.2 **a** System based on optical frequency comb channelization scheme, **b–d** Principle of channelization

spectrally divided by a Fabry–Perot filter with a free spectral range of δ_{FPF}. An optical demultiplexer with a channel spacing of δ_{FPF} is used to divide the spectrum into different channels. The RF spectrum channelization process is shown in Fig. 9.2b–d.

The frequency of the optical frequency comb can be expressed as

$$f_{OFC}(k) = f_{OFC}(1) + (k-1)\delta_{OFC}, \tag{9.6}$$

where f_{OFC} (1) is the frequency of the first root frequency comb. We assume that the Mach–Zehnder modulator is driven by a radio frequency signal with a frequency of f_{RF}, and the modulation method is carrier suppression double-sideband modulation. This radio frequency signal is duplicated by optical frequency comb multicasting, and then the frequency of the kth up-conversion component can be expressed as f_{OFC} (1) + (k-1) δ_{OFC}-f_{RF}. It should be noted that we ignore the + 1 order sideband, because the +1 order sideband cannot pass the Fabry–Perot filter, as shown in Fig. 9.2. For a Fabry–Perot filter, the response function of its lth transmission peak is as follows:

$$S_{FPF}(f, 1) = S_0\{\, f - [f_{FPF}(1) + (l-1)\delta_{FPF}]\}\,, \tag{9.7}$$

where f_{FPF} (1) is the corresponding frequency of the first transmission peak. The spectral pattern $S_0(f)$ of a single transmission peak is as follows:

$$S_0(f) = \frac{(\pi B_{S0})^2}{(\pi B_{S0})^2 + (2\delta_{FPF})^2 \sin^2\left(\frac{\pi f}{\delta_{FPF}}\right)} \left(-\frac{\delta_{FPF}}{2} < f < \frac{\delta_{FPF}}{2}\right), \tag{9.8}$$

where B_{S0} is the full-width at half-maximum of the transmission peak of the Fabry–Perot filter. Note the following: 1) The channel spacing of the optical demultiplexer is the same as the free spectral range of the Fabry–Perot filter and can be accurately aligned; 2) The crosstalk of the optical demultiplexer is negligible; 3) Since the bandwidth of the transmission peak of the Fabry–Perot filter is less than 1 GHz, and the free spectral range of the frequency comb is 39 GHz, only a frequency comb of k = 1 can pass the Fabry–Perot filter. The frequency of the output radio frequency signal is

$$f_{out}(f_{RF}, l) = f_{OFC}(1) + (l - 1)\delta_{OFC} - f_{RF}. \tag{9.9}$$

The intensity is

$$I_{out}(f_{RF}, l) = S_0\{(f_{OFC}(1) - f_{FPF}(1)) - (l - 1)(\delta_{FPF} - \delta_{OFC}) - f_{RF}\}. \tag{9.10}$$

Equations 9.9 and 9.10 describe all characteristics of the channelization scheme. When the input RF signal frequency is $0 < f_{RF} < \delta_{OFC} / 2$ and $B_{S0} << \delta_{OFC}$, we can ignore the +1 order sideband after the Mach–Zehnder modulator. According to Eq. 9.10, the bandwidth of each channel is B_{S0}, which is the same as the bandwidth of $S_0(f)$. It should be noted that we assume $\delta_{FPF} > \delta_{OFC}$, but this solution is also applicable to the case of $\delta_{FPF} < \delta_{OFC}$, but the corresponding formula needs to be modified accordingly. Fabry–Perot filters have a comb-like transmission spectrum, so they can periodically filter specific frequencies. Its free spectral range (FSR) differs from the free spectral range of the optical frequency comb. It is because of the difference in the two free spectral ranges that the *N*th carrier frequency of the optical frequency comb is different from the frequency at the *N*th transmission peak of the Fabry–Perot filter. The Fabry–Perot filter performs comb filtering on multi-frequency RF microwave signals loaded on different carriers of the optical frequency comb. The comb-filtered optical frequency comb passes through the wavelength division multiplexer and is spatially divided into different channels. The power of each channel is detected by a photodetector, thereby inferring the frequency information of the radio frequency microwave signal. This solution can perform real-time, high-speed, parallel and broadband measurements on the RF microwave signals to be measured.

9.1.3 Simulation of Channelized Filter Receiver System

We use simulation to analyze the channelization scheme based on optical frequency comb. In our model, the optical frequency comb ranges from 1551.7 to 1553.8 nm, and the free spectral range is 40 GHz. The 3 dB bandwidth of the Fabry–Perot filter is 400 MHz. We precisely align the optical frequency comb with the transmission peak of the Fabry–Perot filter to ensure $f_{OFC}(1) - f_{FPF}(1) = 13$ GHz. The channel spacing of the optical demultiplexer is 40 GHz. Two RF signals with frequencies of 8 and 10 GHz are modulated on the optical frequency comb. The RF signal is up-converted to the optical frequency and broadcasted through the optical frequency comb. Because the free spectral range of the Fabry–Perot filter is larger than the free spectral range of the optical frequency comb, only the −1 order sideband is used

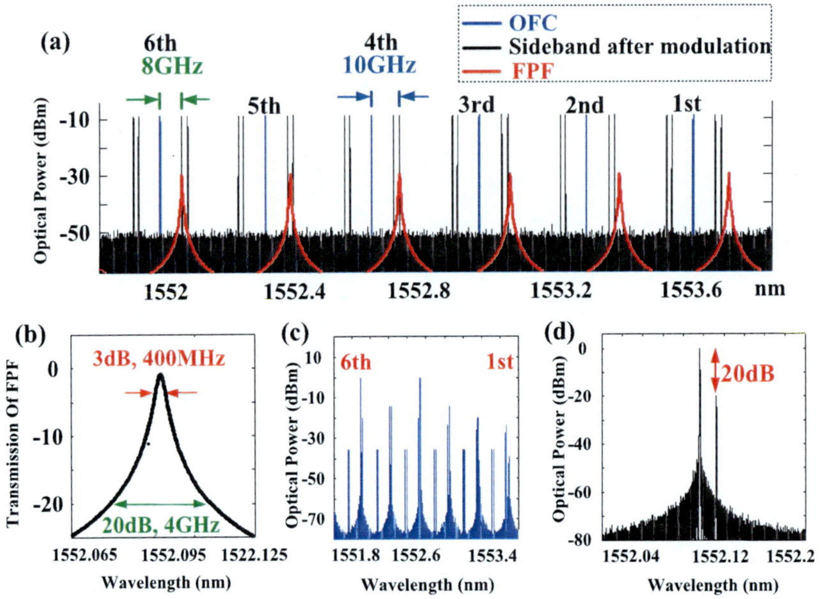

Fig. 9.3 Simulation results: **a** The blue, black and red lines represent the spectrum of the optical frequency comb, the spectrum after modulating the RF signal and the transmission spectrum of the Fabry–Perot filter. **b** Single transmission spectrum of a Fabry–Perot filter. **c** Spectra after passing through a Fabry–Perot filter. **d** Spectrum of the sixth channel output

in the channelization process. As shown in Fig. 9.3a, the 8 GHz signal on the sixth frequency comb is aligned with the sixth transmission peak of the Fabry–Perot filter, while the fourth peak of the Fabry–Perot filter is aligned. The transmission peaks are aligned with the 10 GHz sideband. Figure 9.3c shows the spectrum after passing the Fabry–Perot filter. Optical demultiplexers physically divide the frequency spectrum into different channels. Figure 9.3d shows the spectrum of the sixth channel. It is worth noting that the power difference between the sixth channel and the fifth channel is 14 dB, as shown in Fig. 9.3c. The main reason for this is the Lorentz line of the Fabry–Perot filter. Although the 3 dB bandwidth of the Fabry–Perot filter is 400 MHz, its 20 dB bandwidth is 4 GHz, as shown in Fig. 9.3b. When the sixth transmission peak of the Fabry–Perot filter is aligned with the 8 GHz sideband, we can observe a 20 dB rejection ratio at 10 GHz, as shown in Fig. 9.3d.

9.1.4 Experimental Results and Analysis of Channelized Filter Receiver System

In our experiments, the optical frequency comb was generated by phase modulation using a microwave signal with a frequency of 39 GHz, as shown in Fig. 9.4. The laser

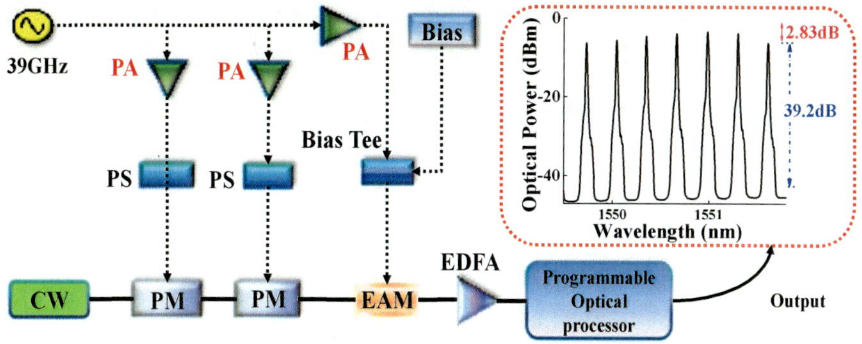

Fig. 9.4 Light frequency generating device. PA: microwave power amplifier; PS: microwave phase shifter; PM: phase modulator; EAM: electric absorption modulator

source we used is a fiber laser (Koheras AdjustiK Benchtop Fiber Laser) with a center wavelength of 1551.67 nm and a line width of less than 1 kHz. Since the number of optical frequency combs is determined by the phase modulation intensity, we load the phase modulator with the highest possible RF power. An electric absorption modulator is then used to planarize the optical frequency comb. At the same time, we need to use a microwave phase shifter to match the two modulators to the microwave signal phase. The optical frequency comb we generated is shown in Fig. 7.4. The frequency comb ranges from 1549.6 to 1551.8 nm. There are seven frequency combs within the 3 dB bandwidth of the optical frequency comb and the signal-to-noise ratio is 40 dB. The maximum detectable RF frequency that this optical frequency comb with a frequency interval of 39 GHz can support is 19.5 GHz.

A Fabry–Perot filter is connected behind the Mach–Zehnder modulator. The 3 dB bandwidth of this filter is 400 MHz. Then we control the filter temperature to align the optical frequency comb with the transmission peak of the filter. In our design, $f_{OFC}(1) - f_{FPF}(1) = 13$ GHz. As a proof of concept, we use a spectrometer to measure the spectrum after a Fabry–Perot filter. The input RF signal is scanned in steps of 1 GHz from 8 to 13 GHz. The measured spectra are shown in Figs. 9.5b and 9.6a–d. We can clearly see that different RF signals appear on different optical frequency combs. We can also observe that due to the Lorentz line of the Fabry–Perot filter, the crosstalk between the channels is 14.4–19.5 dB, which is consistent with our theoretical analysis. In order to prove the channelization of multi-frequency signals, we couple RF signals with frequencies of 8 and 10 GHz and input them into the above-channelized receiver. The spectrum after the Fabry–Perot filter is shown in Fig. 9.6e. We can observe that 8 and 10 GHz RF signals will appear on the fourth and sixth channels, respectively.

As an application of the above-mentioned channelization, we can make instantaneous multi-frequency measurements by directly detecting optical power. Limited to our existing devices, we can calculate the received optical power after filtering, as

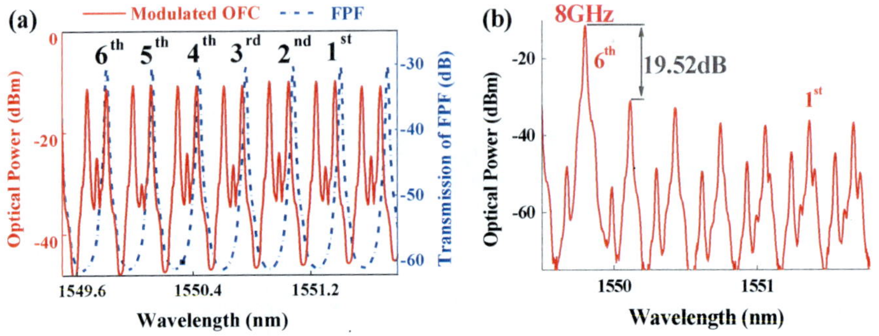

Fig. 9.5 Transmission spectrum of an optical frequency comb and a filter modulated with an 8 GHz RF signal. **b** Spectrum of an optical signal after passing through a Fabry–Perot filter

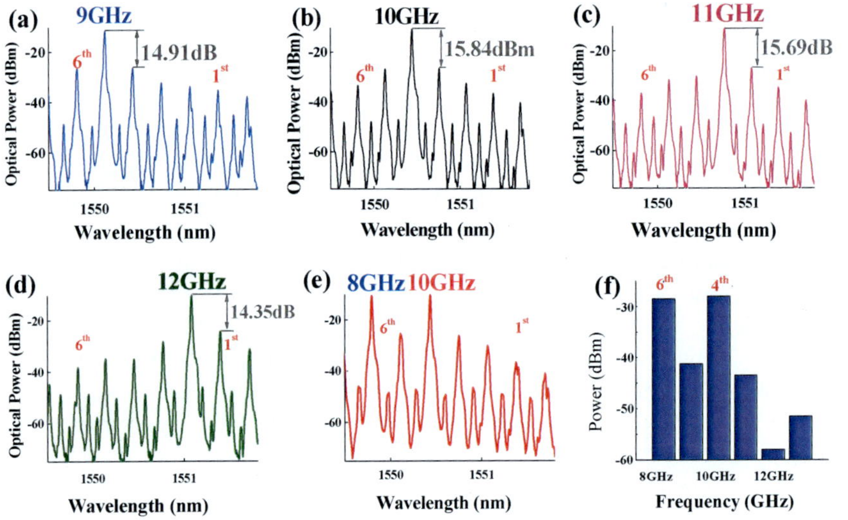

Fig. 9.6 **a–d** At the time when the input RF signal frequency was 9, 10, 11 and 12 GHz, the output spectrum after the Fabry–Perot filter. **e** When a two-tone signal is input, the output spectrum after the Fabry–Perot filter. **f** The power received by each channel

shown in Fig. 9.6f. We assume that the transmission spectrum of the optical demultiplexer is a fourth-order ultra-Gaussian, and the 3 dB bandwidth is 25 GHz. As we have observed, the two-tone signal can be judged in the range of 7.5–8.5 GHz and 9.5–10.5 GHz. Therefore, we have effectively proved the measurement of multitone signals.

9.2 Coherent Optical Communication Technology

In the 1980s, people began to conduct theoretical and experimental research on coherent optical communication. Through research, it has been found that coherent optical communication systems have the advantage of high sensitivity. Therefore, various countries are focusing on developing practical coherent optical communication systems. In 1989, the United States AT&T and Bell Company conducted a non-relay coherent transmission experiment with a 1.7 Gbit/s FSK modulation format at 1300 and 1550 nm wavelengths in Pennsylvania. However, in the 1990s, with the introduction and application of EDFA and WDM technology, the transmission distance and bandwidth of optical fiber communication systems have been greatly improved. Therefore, the development of coherent optical communication technology is relatively slow. However, with the emergence of modern new applications (such as high-definition video, online 3D games) and the widespread application of the Internet, the amount of information in the entire society has shown explosive development. Facing this new requirement, the traditional optical communication system needs to be further improved to better adapt to the great demand for bandwidth in the information age. The traditional optical fiber communication system widely uses the intensity modulation/direct detection method to transmit and receive information. This transmission and reception method has the advantages of simplicity and easy integration. However, this method can only perform envelope detection on signals in the amplitude modulation format. Therefore, the spectrum utilization of this modulation and demodulation technology is very limited, and the bandwidth of a single channel is greatly limited. Therefore, people have begun to study optical communications with advanced modulation formats with higher spectrum utilization. Coherent reception and I/Q demodulation technology are required for the optical carrier signals of the advanced modulation format. The phase information can be obtained through the coherent reception technology, and the demodulation recovery of the signal can be achieved by the I/Q demodulation technology. Below we will briefly introduce coherent optical communication technology.

9.2.1 Coherent Reception Technology

In a coherent optical communication system, at the optical transmitting end, the information to be transmitted is loaded on the optical carrier through various modulation formats through external modulation, and then the optical carrier signal is transmitted through the optical fiber. The detector receives the signal. In a coherent receiving system, the received optical carrier signal is first coherently coupled with a coherent local oscillator signal through an optical coupler. The two output ends of the optical coupler are then connected to the balanced detector, and the coherently coupled optical signals are sent to the balanced detector for detection. According to whether the frequency of the local oscillator signal and the carrier of the received signal are

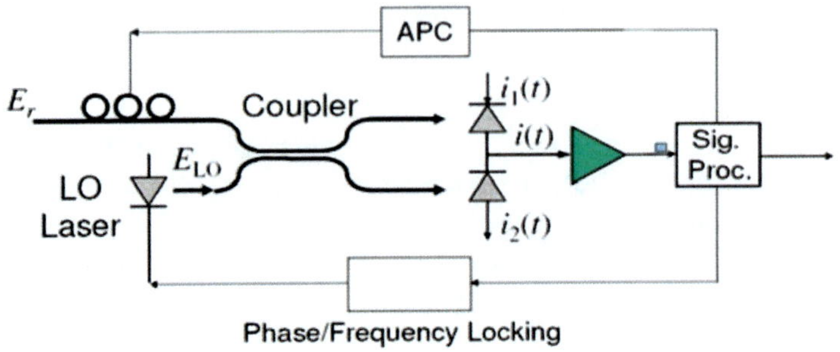

Fig. 9.7 Coherent reception principle

the same, coherent reception can be divided into two receiving modes: homodyne detection and heterodyne detection. The local oscillator light frequency for homodyne detection is the same as the signal light frequency. The optical carrier signal will be directly converted to the baseband after photoelectric conversion. However, this detection method requires that the local oscillator light and the signal light are phase-locked. The degree of balance is also very demanding. The local oscillator optical frequency of the heterodyne detection is different from the signal optical frequency. The optical carrier signal is first down-converted to the intermediate frequency after photoelectric conversion. At this time, the secondary baseband demodulation of the intermediate frequency signal is required.

The principle of coherent reception is shown in Fig. 9.7. The transmission mode of the single-mode fiber is the basic mode HE_{11} mode. The signal light received by the receiver can be expressed by the following formula:

$$E_r = A_r \exp[j(\omega_r t + \varphi_r)], \tag{9.11}$$

where A_r, ω_r and φ_r are the amplitude, frequency and phase of the signal light, respectively. The light field of the local oscillator can be expressed by the following formula:

$$E_{lo} = A_{lo} \exp(j\omega_{lo} t), \tag{9.12}$$

where A_{lo}, ω_{lo} and φ_{lo} are the amplitude, frequency and phase of the local oscillator light, respectively. The polarization controller ensures that the signal light and the local oscillator have the same polarization state. Both the local oscillator light and the signal light are input into a 3 dB 180° optical coupler, then the output light field of the coupler can be expressed as

$$\begin{bmatrix} E_{\text{out,up}} \\ E_{\text{out,low}} \end{bmatrix} = \frac{1}{\sqrt{2}} \begin{bmatrix} 1 & 1 \\ 1 & -1 \end{bmatrix} \begin{bmatrix} E_r \\ E_{lo} \end{bmatrix}. \tag{9.13}$$

From Eq. 7.13 we can get the output light field of the coupler as

$$\begin{aligned} E_{\text{out,up}} &= \frac{1}{\sqrt{2}}(E_r + E_{lo}) \\ E_{\text{out,low}} &= \frac{1}{\sqrt{2}}(E_r - E_{lo}) \end{aligned}. \tag{9.14}$$

The photocurrent signals of the upper and lower arms of the balance detector can be expressed as

$$\begin{aligned} i_1 &= \frac{R}{2}|E_r + E_{lo}|^2 = \frac{R}{2}(A_r^2 + A_{lo}^2) + RA_rA_{lo}\cos[(\omega_{lo} - \omega_r)t - \varphi_s] \\ i_1 &= \frac{R}{2}|E_r - E_{lo}|^2 = \frac{R}{2}(A_r^2 + A_{lo}^2) - RA_rA_{lo}\cos[(\omega_{lo} - \omega_r)t - \varphi_s] \end{aligned}. \tag{9.15}$$

Then after the photoelectric conversion of the balanced detector, we can get the electrical signal down-converted to the intermediate frequency or baseband as

$$i(t) = i_1 - i_2 = 2RA_rA_{lo}\cos[(\omega_{lo} - \omega_r)t - \varphi_s]. \tag{9.16}$$

From the above formula, we can see that the electrical signal after balance detection has no DC component, so coherent reception can suppress common-mode noise. At the same time, the intensity of the converted photocurrent is proportional to the amplitude of the signal light and the amplitude of the local oscillator. Therefore, the intensity of the intermediate frequency signal at the receiving end can be increased by increasing the optical power of the local oscillator. Therefore, coherent reception has higher sensitivity.

9.2.2 I/Q Demodulation Technology

On the basis of coherent reception, by using a 90° optical hybrid coupler and two sets of balanced detectors, it is possible to achieve complete demodulation of the optical carrier signal. The principle of I/Q demodulation technology is shown in Fig. 9.8.

The expressions of signal light and local oscillator light are shown in Eqs. 9.1 and 9.2. Both the local oscillator light and signal light are input into a 90° optical hybrid coupler. The output light field of an ideal optical coupler is

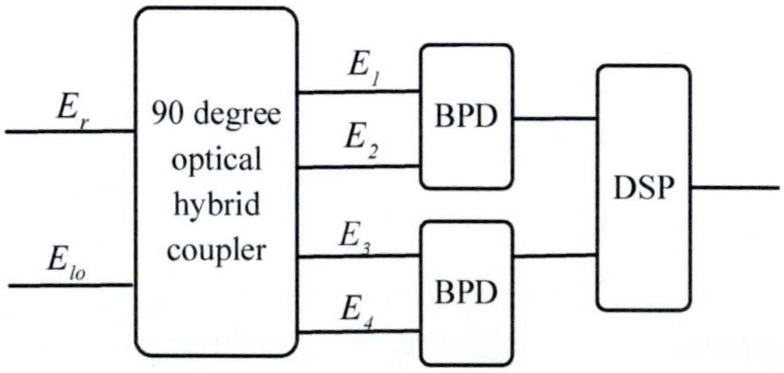

Fig. 9.8 Principle of orthogonal detection

$$\begin{bmatrix} E_1 \\ E_2 \\ E_3 \\ E_4 \end{bmatrix} = \frac{1}{2}\begin{bmatrix} 1 & 1 \\ 1 & -1 \\ 1 & j \\ 1 & -j \end{bmatrix}\begin{bmatrix} E_r \\ E_{lo} \end{bmatrix}. \tag{9.17}$$

From Eq. 9.17 we can get the output light field of the optical hybrid coupler as

$$\begin{array}{l} E_1 = \frac{1}{2}(E_r + E_{lo}) \\ E_2 = \frac{1}{2}(E_r - E_{lo}) \\ E_3 = \frac{1}{2}(E_r + jE_{lo}) \\ E_4 = \frac{1}{2}(E_r - jE_{lo}) \end{array}. \tag{9.18}$$

Then after the photoelectric conversion of the balanced detector, we can get the electrical signals of the I and Q channels down-converted to the intermediate frequency as

$$\begin{array}{l} i_I(t) = RA_rA_{lo}\cos[(\omega_{lo} - \omega_r)t - \varphi_s] \\ i_Q(t) = RA_rA_{lo}\sin[(\omega_{lo} - \omega_r)t - \varphi_s] \end{array}. \tag{9.19}$$

The two signals are then digitized separately, and then the two signals are processed in the digital domain. First, the two signals are integrated into a complex form:

$$i(t) = i_I(t) + j \cdot i_Q(t) = RA_rA_{lo}\exp[(\omega_{lo} - \omega_r)t - \varphi_s]. \tag{9.20}$$

Then strictly phase demodulate the phase information by extracting the phase angle:

$$L = \mathrm{Im}[\ln[i(t)]]. \tag{9.21}$$

Through I/Q demodulation, signals can be flexibly processed in the digital domain to achieve linear demodulation of amplitude-modulated and phase-modulated signals.

9.3 RF Channelization Receiving Technology Based on Dual Coherent Optical Frequency Comb

9.3.1 Coherent Channelization Reception Principle

The proposed channelization scheme is shown in Fig. 9.9. Through the Mach–Zehnder modulator, the signal optical frequency comb (free spectral range is δ_{sig})

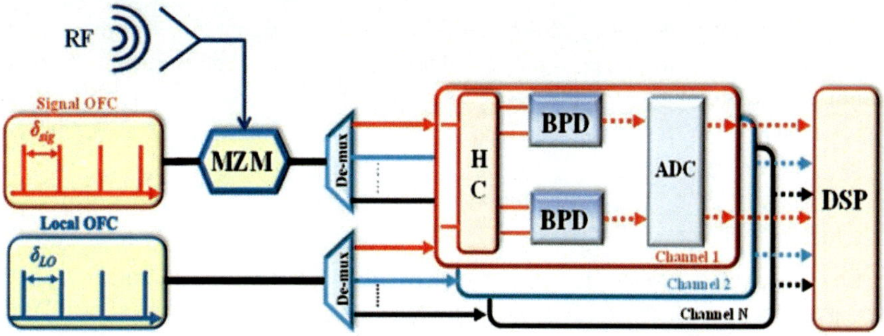

Fig. 9.9 Experimental structure of channelized reception based on coherent optical frequency comb

is modulated by a broadband RF signal. Channel division is then performed by an optical demultiplexer with a channel interval of δ_{lo}. For each channel, the signal optical frequency comb is demodulated by the I/Q demodulation module and the corresponding local oscillator optical frequency comb (the free spectral range is δlo; assuming δ_{lo} and δ_{sig} are slightly different). The standard I/Q demodulation module includes a 90° hybrid coupler (HC), a pair of balanced detectors, an analog-to-digital converter and a digital signal processing module. The broadband RF signal channelization process is shown in Fig. 9.10.

The frequency of each frequency component of the signal optical frequency comb is

$$f_{\text{sig}}(m) = f_{\text{sig}}(1) + (m-1)\delta_{\text{sig}}, \tag{9.22}$$

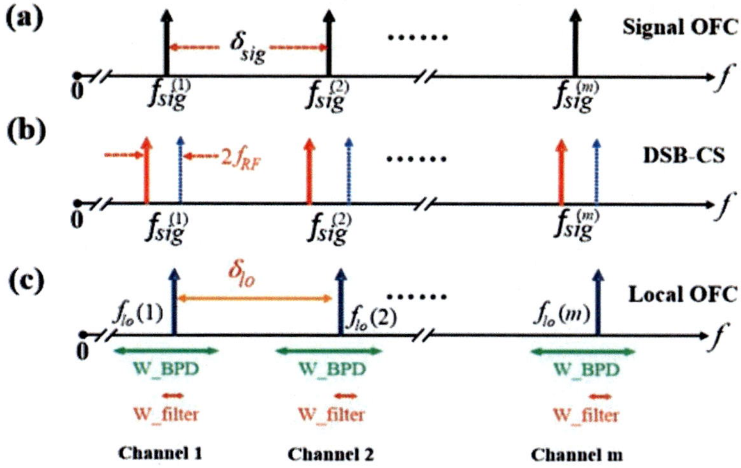

Fig. 9.10 **a** Spectrum of the signal optical frequency comb. **b** The frequency spectrum of the modulated signal light frequency comb. **c** Spectrum of the optical frequency comb of the local oscillator

where $f_{\text{sig}}(1)$ is the frequency of the first frequency component. δ_{sig} and m, respectively, represent the free spectral range and number of signal light frequency combs. Similarly, the frequency of the local oscillator optical frequency comb can be expressed as

$$f_{lo}(n) = f_{lo}(1) + (n - 1)\delta_{lo}, \tag{9.23}$$

where $f_{lo}(1)$ is the frequency of the first local oscillator frequency component. δ_{lo} and n, respectively, represent the free spectral range and number of the local oscillator frequency combs.

We assume that Mach Zendell works under small-signal and carrier-suppressed double-sideband conditions. When a radio frequency signal of frequency f_{RF} is input into the system radio frequency signals are multicast via signal optical frequency combs. The frequency of the mth up-conversion component is

$$f_{\text{sig\_ mod}} = f_{\text{sig}}(1) + (m - 1)\delta_{\text{sig}} + f_{RF}. \tag{9.24}$$

It should be noted that the signal bandwidth obtained by mixing the -1 order sideband with the corresponding local oscillator optical frequency comb exceeds the bandwidth of the digital filter in the balanced detector, analog-to-digital converter or I/Q demodulation module, so the -1 order sideband is ignored. Generally speaking, the channel crosstalk of the optical demultiplexer is very low. Therefore, on the mth channel, the I/Q demodulation module only receives the mth RF replica and the mth local oscillator optical frequency comb, as shown in Fig. 9.10b, c. Two light inputs are able to produce orthogonal amounts of light interference, which are then received by two balanced detectors simultaneously. Channels I and Q represent the real and imaginary parts of the down-converted signal, respectively, and then synthesized in the digital signal processing module. According to Eqs. 9.22 and 9.23, the frequency of the signal received by the mth channel is

$$f_{IF}^{m} = f_{\text{center}}^{m} - f_{RF}, \tag{9.25}$$

where f_{center}^{m} is the center frequency of the mth channel

$$f_{\text{center}}^{m} = [f_{lo}(1) - f_{\text{sig}}(1)] + (m - 1)\Delta, \tag{9.26}$$

where Δ is the channel bandwidth:

$$\Delta = \delta_{lo} - \delta_{\text{sig}}. \tag{9.27}$$

In our scheme, the bandwidth of the balanced detector and the analog-to-digital converter is larger than Δ. In the digital signal processing module, the bandwidth of the digital rectangular filter is Δ and the center frequency is 0. Then the amplitude of the *m*th channel is

$$A_{IF}^{m} = \mathrm{rect}(f_{IF}^{m})\begin{cases} 1 & \left|f_{IF}^{m}\right| < \Delta/2 \\ 0 & \text{others} \end{cases}. \tag{9.28}$$

Since our proposed channelized reception is a linear time-invariant system, Eq. 9.28 characterizes the response of all RF signals loaded with baseband signals through the system. For example, a signal with a bandwidth greater than Δ will be divided into different channels. The output signal of each channel follows Eq. 9.28. It should be noted that we assume $\delta_{\mathrm{sig}} < \delta_{lo}$, but our analysis above is also applicable to the case of $\delta_{\mathrm{sig}} > \delta_{lo}$. Different from the original channelized receiver scheme, in our scheme, the optical demultiplexer is only used to physically divide the broadcast radio frequency channel. In our solution, the channel bandwidth Δ (a few hundred megahertz) of the channelized receiver is much smaller than the channel spacing (δ_{lo}, tens of gigahertz) of the optical demultiplexer. Therefore, the alignment tolerance of the light source and the optical demultiplexer is as high as several gigahertz. This can be achieved with commercially available temperature control modules. Because narrowband filtering and precise frequency shaping can be implemented in digital signal processing modules, our solution avoids the alignment of the light source and the filter. According to Eq. 9.28, we can see that the channelized receiver can achieve seamless coverage of wideband signals, and each channel has an ideal rectangular frequency amplitude response. In our scheme, in order to avoid the −1 order sideband interference of adjacent channels, the frequency of our measurable RF signals is limited by

$$0 < f_{RF} < \delta_{\mathrm{sig}}/2. \tag{9.29}$$

According to Eqs. 9.27 and 9.29, the channelization receiver can support up to the number of channels

$$N < \frac{\delta_{sig}/2}{\Delta}. \tag{9.30}$$

9.3.2 Experimental Structure of RF Channelization Based on Dual-Coherent Optical Frequency Comb

A wideband RF channelization receiving device based on a bi-coherent optical frequency comb and coherent detection. Firstly, two coherent optical frequency combs are generated according to Fig. 9.11. One optical frequency comb is amplified by an erbium-doped fiber amplifier and incident on an intensity modulator that operates in the double-sideband carrier suppression mode to achieve multicasting of the

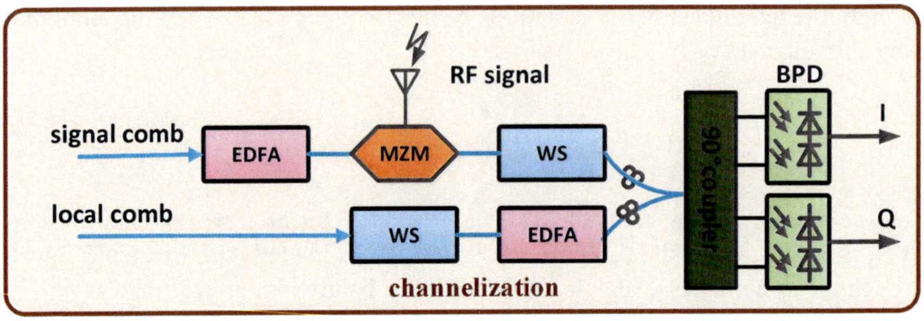

Fig. 9.11 Device of a channelized receiving system based on a coherent optical frequency comb. EDFA: Erbium-doped fiber amplifier; MZM: Mach–Zehnder modulator; WS: Tunable optical filter; BPD: Balanced photodetector

RF signal. An optical filter that filters out a tooth line loaded with a radio frequency signal. The other optical frequency comb filters out the tooth line corresponding to the signal carrier through the optical filter and performs optical amplification. Finally, the two tooth lines are input to the IQ demodulation module for coherent reception. The demodulation module consists of a 90° optical coupler and two balanced photodetectors. Finally, the received signal can be digitally processed and linearly demodulated. Get the RF down-converted signal. In the experiment, a tunable optical filter was used to replace the optical demultiplexer, and a frequency-adjustable radio frequency source was used to replace the broadband radio frequency signal to verify the feasibility of the theory. One channel was used to simulate the multi-channel receiving process.

9.3.3 Experimental Results and Discussion

Firstly, the RF driving frequencies of the optical frequency combs generated in the experiment are 24.5 and 25 GHz, the channel bandwidth is 500 MHz and the frequency range is DC–19.25 GHz to achieve 10 GHz instantaneous receiving bandwidth of 5 GHz. Assume that the continuous optical frequency is f0. The frequency tooth lines and channel parameters of the 10 dual optical frequency combs in DC–4.75 GHz are shown in Table 9.1. Because the tooth line frequency has a symmetrical relationship with the center tooth line on both sides, in order to ensure the beat frequency. The obtained RF local oscillator is in the range of 0–5 GHz, and a frequency difference of 2.5 GHz must be set between the seed light and the signal light. Since the resolution of the optical filter used in the experiment is at least 10 GHz, we will shift it by one cycle here. That is, F0 is set to 27.5 GHz. Correspondingly, F0 is 32.5 GHz in the receiving range of 4.75–9.75 GHz, the center frequency is in the range of 5–9.5 GHz, and the interval is 500 MHz; F0 is 37.5 GHz in the range of 9.75–14.75 GHz, and the center frequency is 10–14.5 GHz; for 14.75–19.75 GHz in

Table 9.1 Channel parameter settings in the 0–5 GHz radio frequency range

Channel	1	2	…	6	…	9	10
Signal OFC(GHz)	f_0-122.5	f_0-98	…	f_0	…	$f_0 + 73.5$	$f_0 + 98$
Local oscillator OFC(GHz)	f_0-122.5	f_0-97.5	…	$f_0 + 2.5$	…	$f_0 + 77.5$	$f_0 + 102.5$
Center frequency (GHz)	0	0.5	…	2.5	…	4	4.5
RF acceptance range (GHz)	0–0.25	0.25–0.75	…	2.25–2.75	…	3.75–4.25	4.25–4.75

the range, F0 is 15.5 GHz, and the center frequency is within 15–19.5 GHz with an interval of 500 MHz.

In addition, the system can also achieve a channel spacing of 1 GHz, using 20 flat wires of a flat optical frequency comb to achieve 20 channels, and the receiving frequency range from DC to 20 GHz. Table 9.2 shows the division of even channels, and the channel parameters such as center frequency, receiving radio frequency range and some test results are given. It can be seen that the receiving errors are in the order of hundreds of kilohertz. The following takes a 20 GHz instantaneous receiving bandwidth system as an example to give a receiver performance test from the aspects of adjacent channel crosstalk, channel equalization and dynamic range. The experiment first tested the channelization function of the system. Adjust the frequency of the RF signal within the range of DC to 20 GHz. All signals fall precisely into the corresponding receiving channel, and all are down-converted to a range of ±500 MHz. The receiving accuracy is within the order of 100 kHz. The received signal has a good signal-to-noise ratio, the average value is greater than 72 dB and the suppression ratio of adjacent channel crosstalk is greater than 50 dB. Figure 9.12 shows the

Table 9.2 Even channel parameters and test results for 1 GHz channel bandwidth

Channel	Medium frequency (GHz)	Acceptance range (GHz)	Test frequency (GHz)	Acceptance frequency (MHz)	Reception error (kHz)
2	1	0.5–1.5	0.693–1.475	307.44–474.48	520
4	3	2.5–3.5	2.606–3.375	394.44–374.72	440
6	5	4.5–5.5	4.618–5.175	381.68–175.20	320
8	7	6.5–7.5	6.973–7.456	26.72–455.92	280
10	9	8.5–9.5	8.946–9.323	53.40–322.52	600
12	11	10.5–11.5	10.618–11.445	381.68–445.48	480
14	13	12.5–13.5	12.875–13.219	125.32–219.28	320
16	15	14.5–15.5	14.937–15.414	62.68–414.16	320
18	17	16.5–17.5	16.947–17.382	53.40–381.68	400
20	19	18.5–19.5	18.549–19.437	451.28–437.36	360

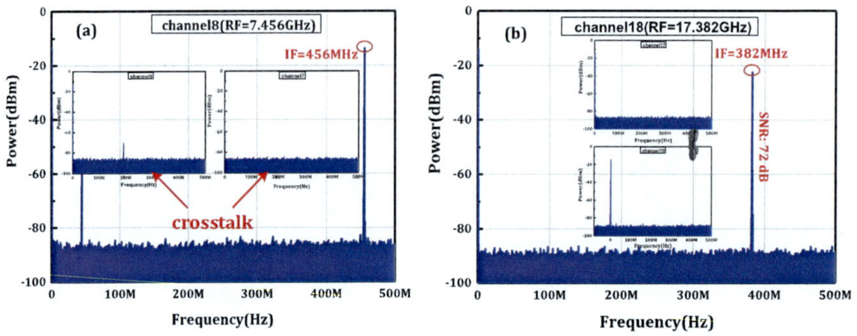

Fig. 9.12 **a** The crosstalk test results of the 8th channel. **b** Test results of adjacent channel crosstalk on the 18th channel

reception results of two optional channels. (A) When the input radio frequency is 7.456 GHz, the corresponding channel is the eighth channel, the center frequency is 7 GHz and the radio frequency is accurately down to 456 MHz. The signal-to-noise ratio is 75 dB, the illustration shows the test results of two adjacent channels under this input, and no adjacent channel crosstalk appears. (B) The test results of the 18th channel when the modulation frequency is 17.382 GHz are the center frequency is 17 GHz, the signal is accurately lower side frequency to 382 MHz, the signal-to-noise ratio is 72 dB and the adjacent channel crosstalk is 52 dB.

Figure 9.13 shows the test results of the frequency response characteristics of the 8th and 18th channels, respectively. The receiving frequency range of the 8th channel is 7.5–8.5 GHz, and the receiving range of the 18th channel is 17.5–18.5 GHz. While maintaining the RF input power unchanged, the frequency is used to test the power of the two channels of received signals in 50 MHz steps. The results show that the channel is well-balanced, the received power jitter is less than 1 dB and the frequency response has an ideal rectangular window shape, ensuring seamless reception of broadband signals.

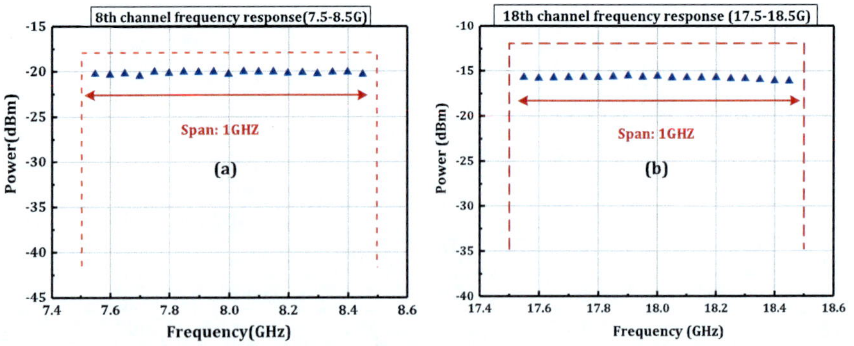

Fig. 9.13 **a** Frequency response test results of the 8th channel. **b** Frequency response test results of the 18th channel

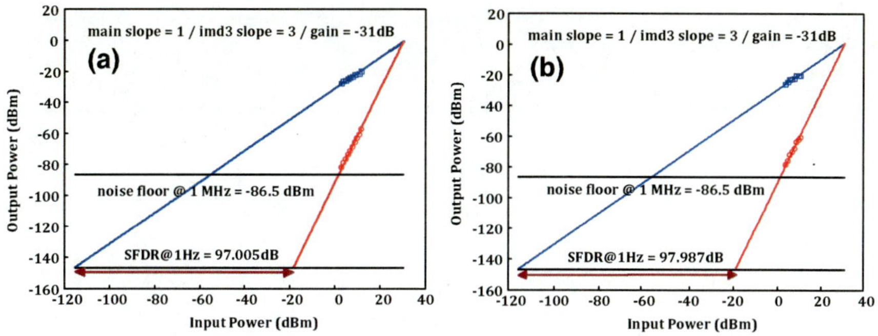

Fig. 9.14 **a** Dynamic range test results of the 8th channel. **b** The dynamic range test results of the 18th channel

Figure 9.14 shows the results of a third-order spurious-free dynamic range (SFDR) experiment with two optional channels. For the test results of the 15th channel, two dual-tone signals 14.875 and 14.895 MHz are selected, and the system is accurately down-converted to 125 and 105 MHz. The corresponding third-order intermodulation distortion is 145 and 85 MHz. Change the power of the input RF from 0 to 10 dBm in increments. Test the power of the output first- and third-order signals to obtain the dynamic range. It can be seen from the figure that the gain of the link is −31 dB and the noise floor is −86.5 dBm/MHz. After testing, the average SFDR of all channels is 97.532 dB $Hz^{2/3}$.

In summary, the system can handle a large signal bandwidth, good flatness of channel frequency response, high signal-to-noise ratio and dynamic range, and is not affected by adjacent channel crosstalk. Beijing Post's RF Photonics Laboratory has for the first time internationally proposed the use of dual-coherent optical frequency combs and I/Q demodulation to achieve the fine-division, down-conversion, and reception integration of broadband radio spectrum, which has attracted widespread international attention. The research laboratory also carried out related projects with CLP 29. Figure 9.15 is the prototype of the receiver we built. With the current integrated development of various core devices, channelized integration is based on optical frequency combs. The system has huge potential in future applications.

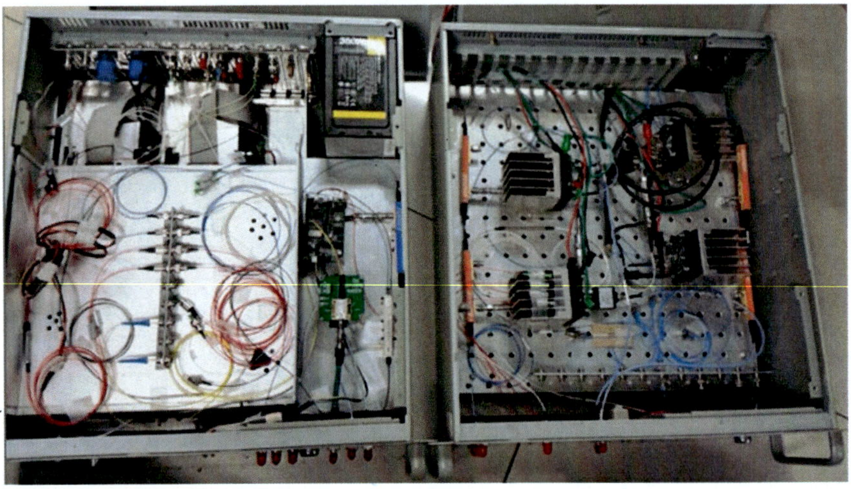

Fig. 9.15 Principle prototype of RF channelization receiver based on dual-coherent optical frequency comb

References

1. S.T. Winnall, A.C. Lindsay, M.W. Austin et al., A microwave channelizer and spectroscope based on an integrated optical Bragg-grating Fabry-Perot and integrated hybrid Fresnel lens system. IEEE Trans. Microw. Tech. **54**(2), 868-872P (2006)
2. D.B. Hunter, R.A. Minasian, Microwave optical filters using in-fiber Bragg grating arrays. IEEE Microwave Guided Wave Lett. **6**(2), 103-105P (1996)
3. P.F. Snawerdt, M.D. Koontz, R.K. Morse et al., Acousto-optic channelizer-based ultra-wideband signal processor, U.S. Patent, 6091522 (2000)
4. C.S. Tsai, Magnetostatic waves-based integrated magnetooptic Bragg cell devices and applications. IEEE Trans. Magnetics. **32**(5), 4118-4123P (1996)
5. A.P. Goutzoulis, Integrated optical channelizer, U.S. Patent. 7421168B1 (2008)
6. X. Zou, W. Pan, B. Luo et al., Photonic approach for multiple-frequency-component measurement using spectrally sliced incoherent source. Opt. Lett. **35**(3), 438-440P (2010)
7. C.S. Brès, S. Zlatanovic, A.O.J. Wiberg et al., Parametric photonic channelized RF receiver. IEEE Photon. Technol. Lett. **23**(6), 344-346P (2011)

Chapter 10
Channelized Link Distortion Compensation Based on Digital Signal Processing

The channelized receiver technology can realize the reconnaissance and reception of multiple signals in a large bandwidth, but the channelized receiver also faces the limitation brought by the system nonlinearity. As we know, the linearity of traditional RF systems is an important indicator. In general, the linearity of an analog optical link is often limited by third-order intermodulation distortion. Many schemes have been proposed to suppress third-order intermodulation distortion in traditional narrowband analog optical links. But for a wideband channelized receiver, the nonlinearity is not limited to the third-order intermodulation distortion, but also the intermodulation distortion between channels [1, 2]. Therefore, the traditional scheme for suppressing third-order intermodulation is not suitable for a wideband channelized receiver. We proposed a method based on the combination of forward distortion signal extraction and backward digital signal processing to achieve simultaneous suppression of third-order intermodulation and interchannel distortion in the channel. With this solution, the channelized receiver can achieve high-fidelity processing and reception of wideband multi-signals.

10.1 Significance of Channelized Link Distortion Compensation

Due to a large number of modern military electronic equipment, the electromagnetic environment is becoming increasingly complex, and mutual interference is serious. As a result, the receiver is receiving more and more interference when receiving signals, and the ability to receive signals is becoming weaker. How to improve the sensitivity and selectivity of the receiver has become an increasingly important research direction in modern RF system design. Modern RF receivers are mainly developed toward high frequency, high linearity, large dynamic range, high sensitivity and high resolution. With the rapid development of modern modulation systems and the increasing utilization of wireless spectrum, the performance and

J. Zhang and J. Li, *Satellite Photoelectric Sensing Technology*,
https://doi.org/10.1007/978-3-030-89843-4_10

indicators of receiver linearity, dynamic range, sensitivity, anti-interference ability and adaptability have become increasingly demanding. This requires the receiver to improve its linearity as much as possible under the premise of ensuring the signal detection capability (high sensitivity) so as to minimize signal distortion and bit error rate; at the same time, widen the dynamic range of the receiver as much as possible to make it adaptable larger and stronger anti-interference ability. Specifically, the radar receiver must have a frequency coverage of 0–18 GHz or even higher, a dynamic range greater than 90 dB $Hz^{2/3}$ and a sensitivity less than − 100 dBm [3, 4].

In modern military equipment, down-conversion modules have been widely used to facilitate digitization and digital processing in signal processing units. The down-conversion module has been widely used in fixed broadband wireless access, wireless local loop, personal mobile communication equipment, military equipment, microwave links and digital spread spectrum communication systems. In a typical application of a traditional receiver, a method of mixing locally generated oscillation waves with an input signal to transform the frequency of the input signal into a predetermined frequency is used, and the RF-modulated signal is converted to an easily processed intermediate frequency. IF-modulated signal is amplified, filtered, demodulated and processed by DSP. By utilizing the powerful and flexible processing capabilities of the DSP, the functions of modern military electronic equipment have been greatly improved. However, due to the sampling rate of the ADC, it is still difficult for modern military electronic equipment to process multiple broadband signals that arrive at the same time in real time [5–7]. Through the introduction in Chaps. 4 and 5, the above goals can be achieved by using photon channelized reception. Now we need to further analyze the main performance indicators of channelized reception. The nonlinear distortion of the signal has a serious impact on the system, and various schemes are needed to suppress the nonlinear distortion in the link. However, the traditional distortion suppression scheme is only for narrowband systems. Therefore, we propose a channelized link distortion compensation technology based on digital signal processing to achieve high-fidelity reception and processing of signals by channelized reception.

10.2 Nonlinear Distortion Generation Mechanism in Multi-Carrier RF Optical Links

Affected by the nonlinearity of various devices, the optical link can be regarded as a nonlinear two-port network. For the system that directly detects the intensity modulation, the nonlinearity of the modulator is mainly considered. When the radio frequency signal $v_1(t) = V_0 \cos(\omega_1 t)$ and the two-tone signal $v_2(t) = V_0[cos(\omega_1 t) + cos(\omega_2 t)]$ are substituted into formula (7.1) and subjected to the first-type n-order Bessel function expansion, the output electrical signals are:

$$I_1(t) \approx I_{dc}\left\{\begin{array}{l} 2\sin\Delta\phi \cdot J_1(\beta)cos(\omega_1 t) \\ -2\sin\Delta\phi \cdot \sum_{n=1}^{\infty} J_{2n+1}(\beta)cos[(2n+1)\omega_1 t \\ +2\cos\Delta\phi \cdot \sum_{n=1}^{\infty} J_{2n}(\beta)cos(2n\omega_1 t) \end{array}\right\} \quad (10.1)$$

$$I_2(t) = 2I_{dc}\cdot\sin\Delta\phi\cdot\left\{\begin{array}{l} J_0(\beta)\sum_{m=1}^{\infty} J_{2m+1}(\beta)sin[(2m+1)\omega_1 t] \\ +J_0(\beta)\sum_{n=1}^{\infty} J_{2n+1}(\beta)sin[(2n+1)\omega_2 t] \\ +\sum_{m=0}^{\infty}\sum_{n=1}^{\infty} J_{2m+1}(\beta)J_{2n}(\beta)sin[2n\omega_2 t \pm (2m+1)\omega_1 t] \\ +\sum_{m=0}^{\infty}\sum_{n=1}^{\infty} J_{2n+1}(\beta)J_{2m}(\beta)sin[2m\omega_1 t \pm (2n+1)\omega_2 t] \end{array}\right\} \quad (10.2)$$

where β is the modulation depth of the modulator. As can be seen from the above two formulas, in addition to the fundamental frequency signal, the output also contains a high-order nonlinear distortion term, which manifests a lot of clutter on the spectrum. Accurate spectrum analysis produces severe interference. Under the input of a wideband multi-carrier radio frequency signal, the input signal is represented as $v(t) = \sum_k v_k(t)\cos[\omega_k t + \varphi_k(t)]$, where $v_k(t)$ is the amplitude, ω_k of each carrier is the angular frequency and $\varphi_k(t)$ is the phase noise of each carrier, which substituted in (10.1) can get the kth output:

$$S_k(t) \propto \prod_p J_0(\beta v_p(t)) \cdot \frac{J_1(\beta v_k(t))}{J_0(\beta v_k(t))} \cdot \sin[\omega_{IF} t + \varphi_k(t)]. \quad (10.3)$$

The first term on the right side of the above formula is the amplitude of the kth carrier. It can be seen that its amplitude is related to the amplitude of all other carriers, that is, the intermodulation distortion between carriers. The second term on the right is only related to itself. It describes the intermodulation distortion. It can be seen that under multi-carrier input, each carrier will affect each other. Intermodulation distortion and inter-carrier distortion coexist, as shown in Fig. 10.1. The suppression of the second harmonic can be changed by adjusting the bias voltage of the modulator. The working point is used to realize that the third and higher harmonics can be filtered by filtering because they are far away from the baseband signal. In-band third-order

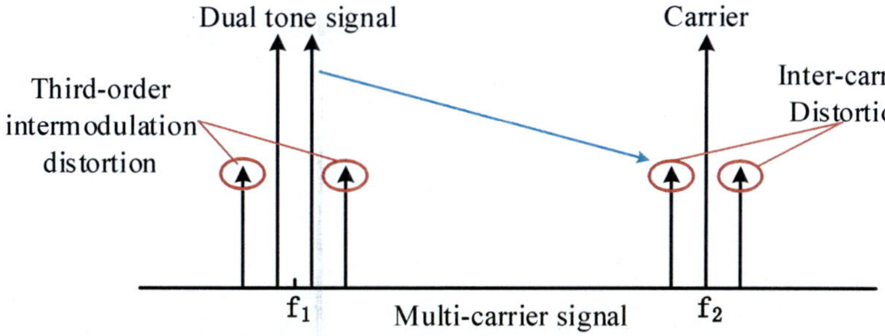

Fig. 10.1 Intermodulation distortion and inter-carrier distortion between multi-carrier signals

intermodulation distortion and inter-channel distortion between carriers need to be linearized to suppress and compensate because they are closer to the target signal.

10.3 Channelized Link Distortion Compensation Principle Based on Digital Signal Processing

We propose a method based on the combination of forward distortion signal extraction and backward digital signal processing to achieve simultaneous suppression of third-order intermodulation and inter-channel intermodulation distortion in the channel. With this solution, the channelized receiver can achieve high-fidelity processing and reception of wideband multi-signals. The principle of this solution is shown in Fig. 10.2.

As shown in Fig. 10.2, the light output by the laser is modulated by a Mach–Zehnder modulator onto a broadband RF signal. The RF signal can be expressed as

$$V(t) = \sum_k v_k(t) \sin[\omega_k t + \varphi_k(t)], \tag{10.4}$$

where $v_k(t)$ is the amplitude envelope of the signal and $\varphi_k(t)$ is the phase of the signal. The Mach–Zehnder modulator is biased at a low bias point, so the upper output arm of the Mach–Zehnder modulator is in a state where the carrier passes, and the lower arm is in a state of carrier suppression. The lower arm performs balance detection with the local oscillator to obtain the output IF signal

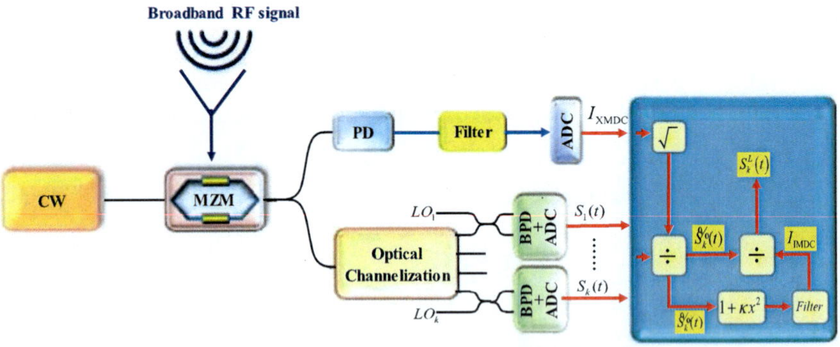

Fig. 10.2 Scheme schematic diagram, where CW is continuous light, MZM is a Mach–Zehnder modulator, BPD is a balanced detector and ADC (analog digit converter) is an analog-to-digital converter

$$S_k(t) \propto \prod_p J_0\left[\beta v_p(t)\right] \times \frac{J_1[\beta v_k(t)]}{J_0[\beta v_k(t)]} \times \sin[\omega_{IF} t + \varphi_k(t)], \tag{10.5}$$

where $\beta = \pi/V_\pi$ (V_π is the half-wave voltage of the modulator), $J_n(x)$ is the nth order Bessel function, ω_{IF} is the frequency of the intermediate frequency signal. The first term in the above formula represents the intermodulation distortion between channels, and the second term represents the intermodulation distortion within the channel. The upper arm of the Mach–Zehnder modulator contains inter-channel intermodulation distortion information, so the optical signal is received through the PD to provide inter-channel intermodulation distortion information for subsequent digital signal processing:

$$I_{XMDC}(t) \propto \left\{\prod_p J_0\left[\beta v_p(t)\right]\right\}^2. \tag{10.6}$$

This signal is a low-frequency signal, which can be analog-to-digital converted by a high-precision ADC, and then compensate the channel intermodulation distortion in the digital domain:

$$\begin{aligned}\tilde{S}_k(t) &= \frac{S_k(t)}{\sqrt{I_{XMDC}}} \propto \frac{J_1[\beta v_k(t)]}{J_0[\beta v_k(t)]} \sin[\omega_{IF} t + \varphi_k(t)] \\ &\approx \frac{1}{2}\left\{1 + 0.125[\beta v_k(t)]^2\right\}\beta v_k(t) \sin[\omega_{IF} t + \varphi_k(t)].\end{aligned} \tag{10.7}$$

Then extract the intermodulation distortion in the channel by digital square sum filtering:

$$I_{IMDC}(t) = 1 + 0.125(\beta v_k)^2 \approx 1 + \kappa F\left\{\left[\tilde{S}_k(t)\right]^2\right\}. \tag{10.8}$$

The output IF signal after the final linearization is

$$S_k^L(t) = \tilde{S}_k(t) \Big/ I_{IMDC}(t) = \frac{S_k(t)}{I_{IMDC}(t)\sqrt{I_{XMDC}}}. \tag{10.9}$$

With this solution, inter-channel intermodulation distortion existing in a wideband channelized receiver can be simultaneously compensated, thereby realizing the high-fidelity reception and processing of wideband high-frequency microwave signals.

10.4 Nonlinear Suppression Experiment and Performance Discussion of Channelized Link Based on Digital Processing

In the experiment, a channel is selected for performance testing of the compensation algorithm. The experimental scheme is shown in Fig. 10.3. Firstly, a continuous beam of light is divided into two paths by an optical coupler. The modulator modulates a radio frequency local oscillator and then uses an optical filter to filter out a sideband to act as an optical local oscillator. On the road, a dual-drive intensity modulator is used to load the RF signal, and the drive voltage is modulated to work at a low bias point. Then, the road is used to extract the distortion compensation signal $I_{XMDC}(t)$, and the road is operated under the bilateral band carrier suppression state, as the modulation signal and optical vibration is down-converted to obtain the baseband signal $S_k(t)$. Due to the use of coherent detection, in order to cancel the additive noise of the link as much as possible, we have performed delay matching on the two signals. In addition, the degree of phase matching between the compensation signal and the baseband signal will also have a greater impact on subsequent distortion compensation effects, so the phase matching between the two signals is also essential. Finally, the real-time sampling filter is used to convert the compensation signal and the baseband signal into digital signals for algorithm processing. Assume that the modulated RF signal is two-tone signals with a long center frequency separation:

$$v(t) = A_1 \cos(2\pi\delta_1 t)cos(2\pi f_1 t) + A_2 \cos(2\pi\delta_2 t)cos(2\pi f_2 t), \tag{10.10}$$

where A_1 and A_2 are signal amplitudes, f_1 and f_2 are the center frequencies of the two-tone signals, and δ_1 and δ_2 are the frequency intervals of the two-tone signals. A two-tone signal with a center frequency of f_1 is down-converted as a baseband signal, and the other two-tone signal has a higher carrier and belongs to an out-of-band signal.

When δ_1 is 7 MHz, the corresponding dual-tone signals are 15 and 15.007 GHz, and they are down-converted to about 80 MHz after coherent detection. The double-tone signals corresponding to 1.5 MHz of δ_2 are 3 and 3.0015 GHz, and the corresponding received signal spectrum structure is shown in Fig. 10.4. The blue line in the figure is the spectrum result before nonlinear suppression. The power ratio between the third-order intermodulation and the signal is 52 dB, and the intermodulation distortion between the carriers and the signal power is 44 dB. The modulation is suppressed by 19 dB and the intermodulation distortion is suppressed by 23.6 dB, as shown by the red dotted line. It can be seen that the algorithm has obvious suppression of intermodulation distortion in-band and intermodulation distortion between out-of-band carriers. The magnitude of the rejection ratio mainly depends on the accuracy of the phase matching between the compensation signal and the baseband signal, and the deviation between the value of the compensation coefficient k and the theoretical value in the formula (10.8).

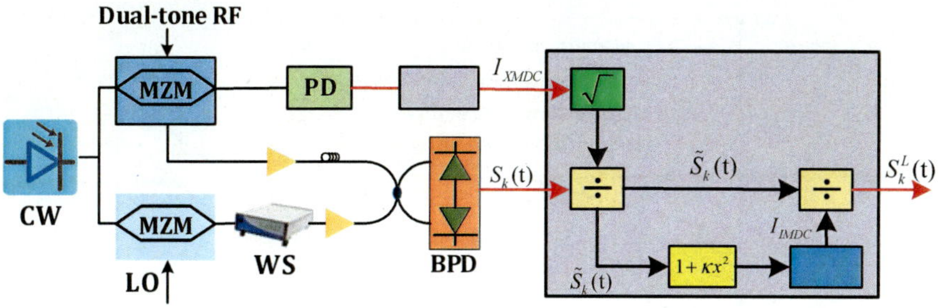

Fig. 10.3 Experimental device for nonlinear suppression of optical link based on digital post compensation. WS: Tunable optical filter

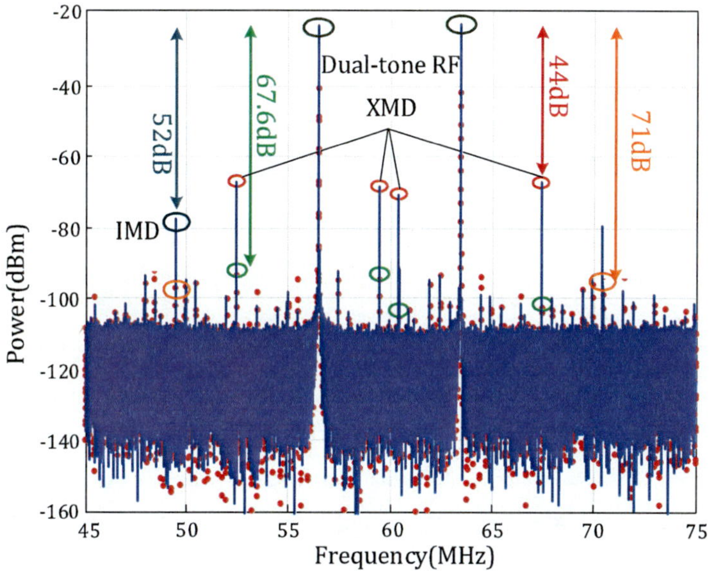

Fig. 10.4 Nonlinear suppression results of optical link based on digital post-compensation

At the same time, we calculated the dynamic range of the channel based on the digital post-compensation algorithm. The results are shown in Fig. 10.5. The third-order spurious-free dynamic range before compensation is 109 dB $Hz^{2/3}$. The range has been increased to 117 dB $Hz^{2/3}$, which greatly increases the channel's ability to handle strong and weak signals.

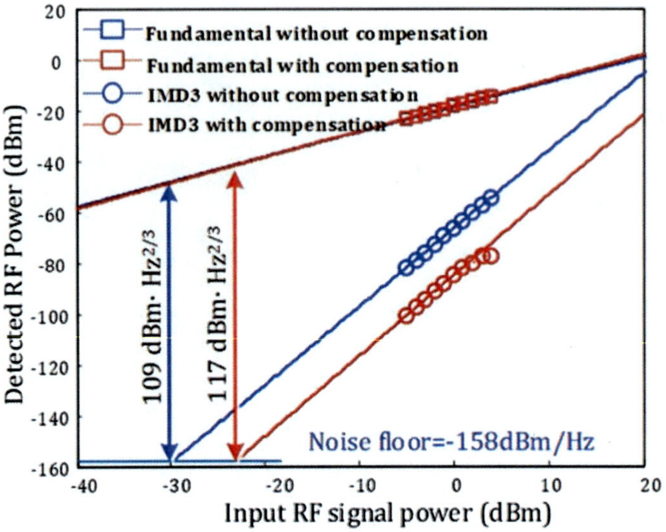

Fig. 10.5 Measurement results of the digital post-compensation algorithm on channel dynamic range

References

1. P.D. Hale, F.V. Kowalski, Output characterization of a frequency shifted feedback laser: theory and experiment. J. Quantum Electron. **26**(10), 1845–1851 (1990)
2. H. Guillet de Chatellus, O. Jacquin, O. Hugon et al., Generation of ultrahigh and tunable repetition rates in CW injection-seeded frequency-shifted feedback lasers. Optical Soc. Am. **21**(13), 15065–15074 (2013)
3. T. Jannson, Real-time Fourier transformation in dispersive optical fibers. Opt. Lett. **8**(4), 232–234 (1983)
4. Y. Park, T.-J. Ahn, J.-C. Kieffer et al., Optical frequency domain reflectometry based on real-time Fourier transformation. Opt. Express **15**(8), 4597–4616 (2007)
5. K. Goda, B. Jalali, Dispersive Fourier transformation for fast continuous single-shot measurements. Nat. Photonics **7**(2), 102–112 (2013)
6. H.G. De Chatellus, L.R. Cortés, J. Azaña, Optical real-time Fourier transformation with kilohertz resolutions. Optica **3**(1), 1–8 (2016)
7. C. Wang, J. Yao, Chirped microwave pulse generation based on optical spectral shaping and wavelength-to-time mapping using a sagnac loop mirror incorporating a chirped fiber bragg grating. J. Lightwave Technol. **27**(16), 3336–3341 (2009)